KU-012-617

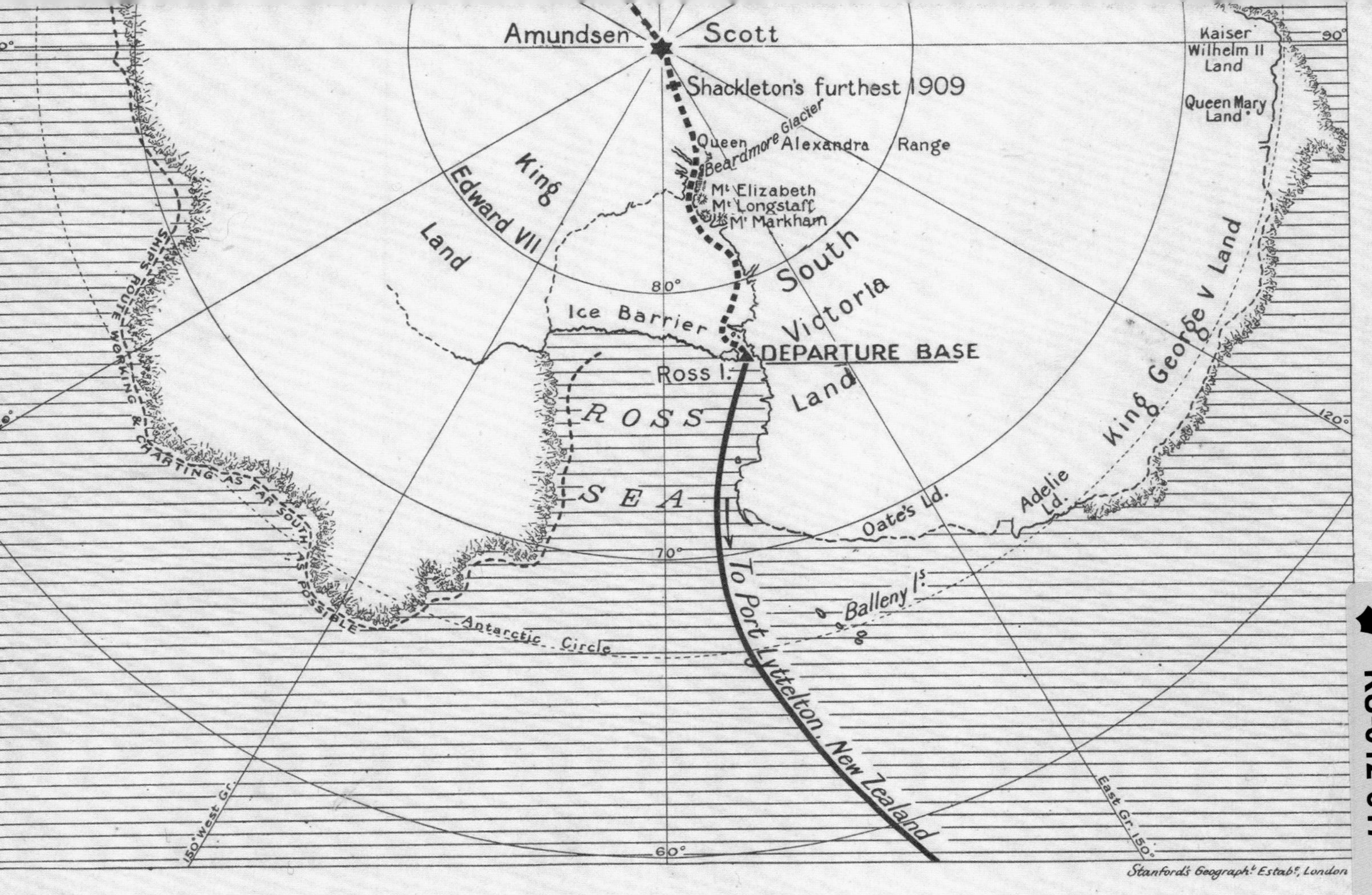
Amundsen
Scott
Shackleton's furthest 1909
Queen Alexandra Range
Beardmore Glacier
Mt. Elizabeth
Mt. Longstaff
Mt. Markham
King Edward VII Land
South Victoria Land
Ice Barrier
DEPARTURE BASE
Ross I.
ROSS SEA
Oate's Ld.
Adelie Ld.
King George V Land
Kaiser Wilhelm II Land
Queen Mary Land
Balleny Is.
To Port Lyttelton, New Zealand
Antarctic Circle
SHIPS ROUTE WORKING & CHARTING AS FAR SOUTH AS POSSIBLE
150° West Gr.
East Gr. 150°
90°
120°
80°
70°
60°
Stanford's Geograph. Estab., London

Shackleton

Exclusively Signed

Ran Fiennes

RANULPH FIENNES

Shackleton

A Biography

RANULPH FIENNES

MICHAEL PENGUIN MJ Est. 1935 JOSEPH

MICHAEL JOSEPH

UK | USA | Canada | Ireland | Australia
India | New Zealand | South Africa

Michael Joseph is part of the Penguin Random House group of companies
whose addresses can be found at global.penguinrandomhouse.com

First published 2021
001

See page 383 for picture credits

Set in 13.5/16 pt Garamond MT Std
Typeset by Jouve (UK), Milton Keynes
Printed and bound in Great Britain by Clays Ltd, Elcograf S.p.A.

The authorized representative in the EEA is Penguin Random House Ireland,
Morrison Chambers, 32 Nassau Street, Dublin D02 YH68

A CIP catalogue record for this book is available from the British Library

HARDBACK ISBN: 978–0–241–35671–5
TRADE PAPERBACK ISBN: 978–0–241–35672–2

www.greenpenguin.co.uk

This book is dedicated to the sponsors and team members of the expeditions of Ernest Shackleton as well as those of my own.

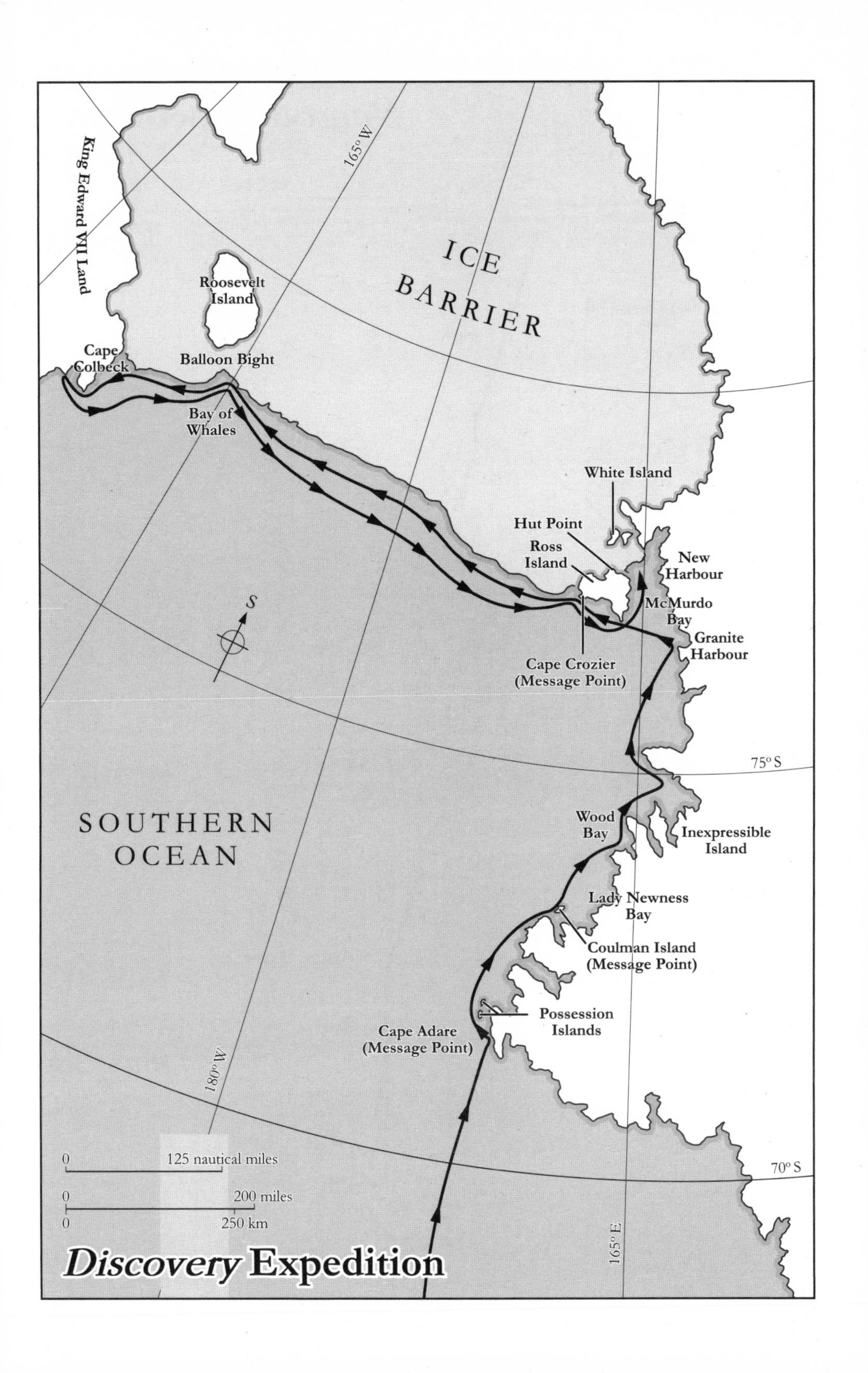

Discovery Expedition

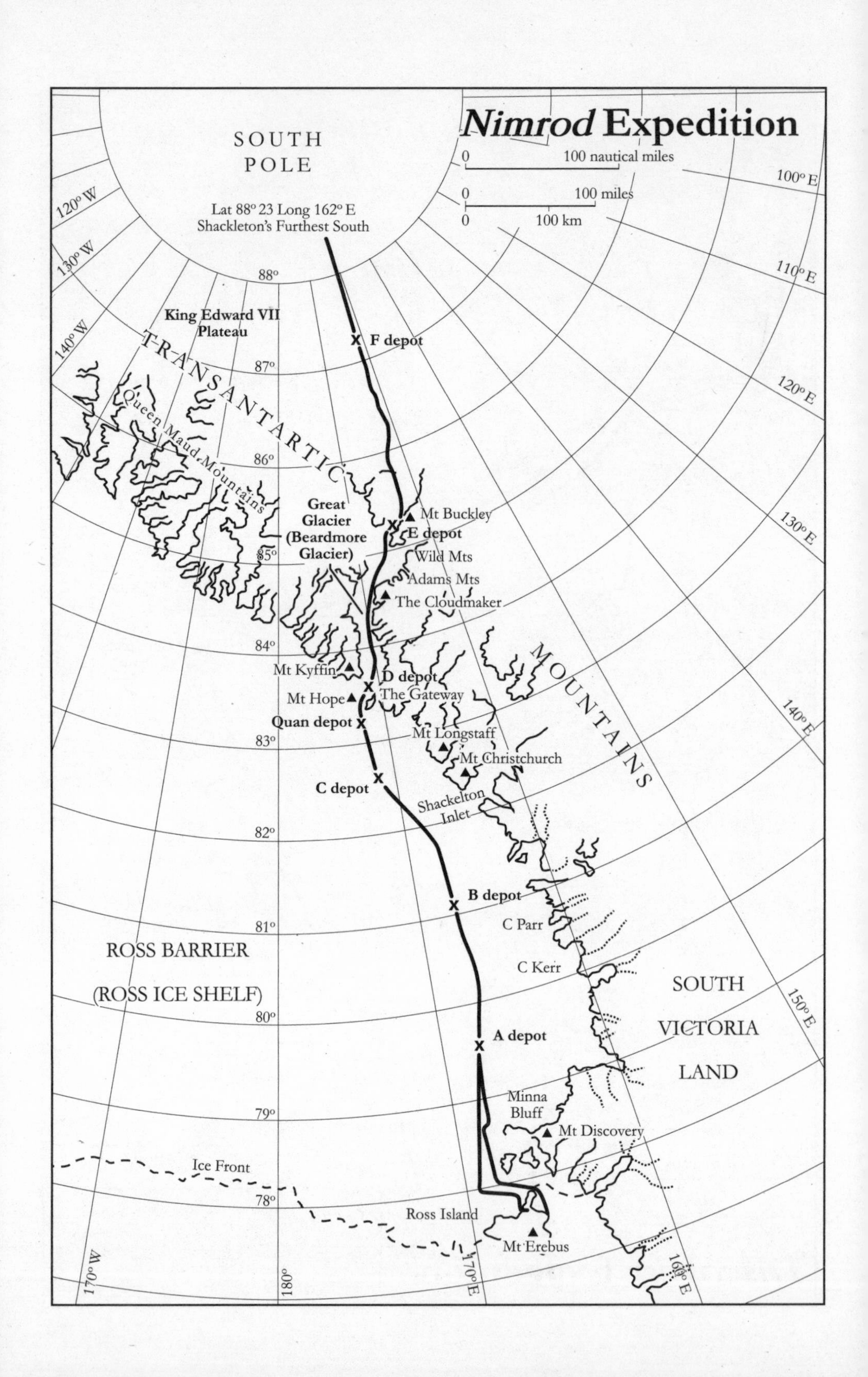

Nimrod Expedition
0
100 nautical miles
0
100 miles
0
100 km
SOUTH
POLE
Lat 88° 23 Long 162° E
Shackleton's Furthest South
120° W
130° W
140° W
170° W
180°
170° E
160° E
150° E
140° E
130° E
120° E
110° E
100° E
88°
87°
86°
85°
84°
83°
82°
81°
80°
79°
78°
King Edward VII
Plateau
TRANSANTARTIC
MOUNTAINS
Queen Maud Mountains
F depot
Great
Glacier
(Beardmore
Glacier)
Mt Buckley
E depot
Wild Mts
Adams Mts
The Cloudmaker
Mt Kyffin
D depot
Mt Hope
The Gateway
Quan depot
Mt Longstaff
Mt Christchurch
C depot
Shackelton
Inlet
B depot
C Parr
C Kerr
ROSS BARRIER
(ROSS ICE SHELF)
SOUTH
VICTORIA
LAND
A depot
Minna
Bluff
Mt Discovery
Ice Front
Ross Island
Mt Erebus

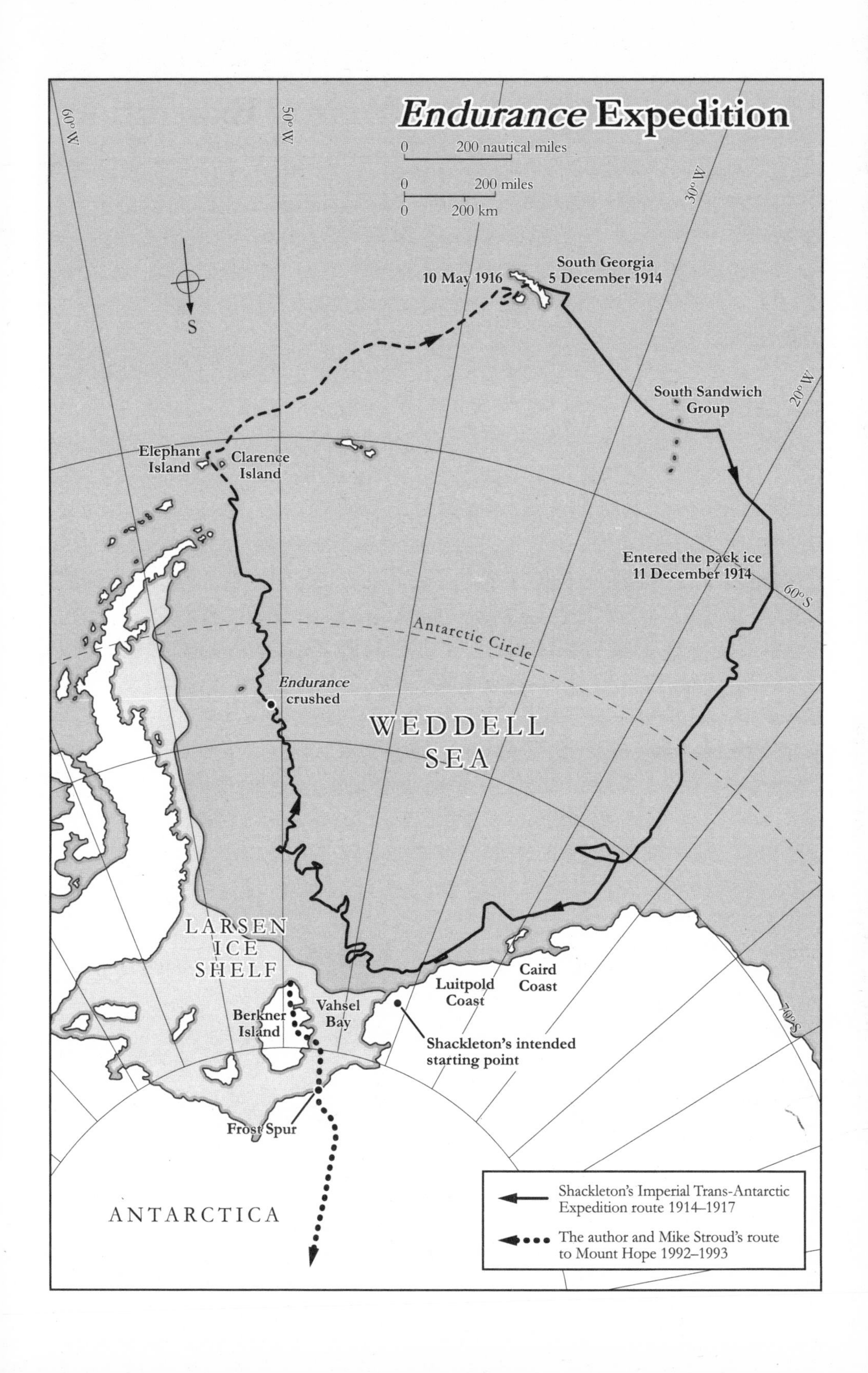

Endurance Expedition
0 200 nautical miles
0 200 miles
0 200 km
60°W
50°W
30°W
20°W
60°S
70°S
S
South Georgia
5 December 1914
10 May 1916
South Sandwich Group
Elephant Island
Clarence Island
Entered the pack ice
11 December 1914
Antarctic Circle
Endurance crushed
WEDDELL SEA
LARSEN ICE SHELF
Berkner Island
Vahsel Bay
Luitpold Coast
Caird Coast
Shackleton's intended starting point
Frost Spur
ANTARCTICA
Shackleton's Imperial Trans-Antarctic Expedition route 1914–1917
The author and Mike Stroud's route to Mount Hope 1992–1993

Introduction

During the searing-hot summer of 1964, I was packed off to London to purchase canoes on behalf of the Royal Scots Grey's Canoe Club. The so-called 'swinging sixties' were well underway, so I needed no second invitation to escape my barracks and visit a city bustling with life and excitement.

Having hailed a black cab from Waterloo Station, as we passed the shimmering Thames I noticed quite a commotion taking place on board a ship. Arching my neck to get a better look, the black-cab driver shouted over his shoulder, 'Them's all Shackleton men . . . famous explorers and all that.'

Shackleton . . . The mere mention of his name conjured up a buzz within me. I knew the stories well, having been enthralled by them as a child. An explorer and conqueror of unknown lands, Shackleton was also the man who had embarked on one of the most dramatic survival and rescue missions of all time. While his name was synonymous with the Antarctic, it was also indelibly linked with virtues such as courage and leadership. Once a mere mortal, Shackleton was now an icon, whose endeavours were forever etched into the granite of British history.

Forty-two years after his death, a celebration of his life was now being held on the Thames, the scene of departure for so many of his expeditions, where he had set off in front of thousands of cheering admirers. Time, it seemed, had not dulled the Shackleton legend. I was later to learn that on the ship that night were fellow explorers, journalists and admirers, as well as the two doctors who were with him when he died,

Alexander Macklin and James McIlroy. It seemed everyone still wanted to somehow touch the magic, which despite Shackleton's many failings, and failures, had made him the most colourful of all the famous Victorian polar tyros.

Within seconds of setting my eyes on this scene we had passed the ship and I had to return my mind to the less exciting job at hand: buying canoes. But in its own way this seemingly inconsequential moment was already laying a path for me to follow in years to come.

Having left the armed forces in the early seventies, and in need of a job, I embarked on some expeditions along various remote rivers and glaciers in Norway and Canada. My appetite whetted, and my ambition and confidence gathering, I then set my mind to conquering some of the great polar challenges which had eluded the likes of Scott, Shackleton and even the famed Norwegian Roald Amundsen.

The first of these was the Trans Globe Expedition. After seven years of excruciating fundraising and preparations, my team and I set off in 1979, with the aim of becoming the first to travel over the entire surface of the Earth, via both poles, without flying one yard of the way. This, of course, not only involved reaching the South Pole but travelling across Antarctica itself, as had once been Shackleton's dream. In doing so, much like the famous Victorian explorers, I was able to set sight on lands that had never before been seen, let alone crossed. Like Shackleton and his contemporaries, we also did this before the age of polar-orbiting satellites, so, with no GPS, satnav or satphone, we had to navigate with sextants, theodolites and hand compasses.

In 1993, I went one step further, when along with my expedition colleague, Mike Stroud, we became the first to cross the Antarctic continent unassisted, without the aid of food drops, transport or machinery. For ninety-two days, we each hauled a load of 485 lbs, over snow, ice, treacherous crevasses and mountains. All the while, we fought starvation, fatigue, frostbite and blindness, as well as the constant danger of falling into the crevasses. In 1996, with most of our polar goals now accomplished, I attempted to become the first man to cross the Antarctic solo and unassisted. Alas, an attack of kidney stones stopped me short.

Nevertheless, these expeditions have provided me with a unique perspective into Shackleton the man, as well as his adventures. As the Shackleton legend has grown over the decades, a litany of books and films about him has flooded the market. Debates have subsequently raged about his preparations and decision-making, not all of them flattering. Some of these books have certainly been entertaining, and educational, but others I have found myself vehemently disagreeing with, as some have played many tunes, invented many twists and told many lies.

It is for that reason that I decided to write this book. Of course, I needed no second invitation to travel alongside Shackleton and live vicariously through his adventures. But I certainly don't want to simply retread familiar tales. My aim is to offer my own perspective, for what it's worth, to hopefully enlighten and enrich the legend. Indeed, to write about Hell, it certainly helps if you have been there, and no previous Shackleton biographer has man-hauled a heavy sledge load through the great crevasse fields of the Beardmore Glacier, explored undiscovered icefields or walked a thousand miles on poisoned feet, hundreds of miles away from civilization.

I hope you will enjoy Shackleton's still quite incredible story, for even now, almost a century after his death, his great achievements continue to astound and amaze.

PART ONE

'And I smelt the Galley's odour
Heard curses of sailor men'

I

The giggling, cries and gasps of awe coming from the young Ernest Shackleton's bedroom were signs that, once again, he had his sisters in the palm of his hand. Gathering around their brother, the Shackleton sisters, of whom there would eventually be eight, were totally immersed in his grip. Standing tall, looking at his sisters one by one, the young boy, with fair hair and angelic eyes, set forth tales from his vivid imagination. He told them that just weeks before he had gone to London with one of his friends and they had encountered a raging inferno which threatened to engulf the city. Together, they had somehow managed to save the day and, as a reward, the Monument, near London Bridge, had been erected in their honour.

Despite such an outlandish tale, the Shackleton sisters believed it to be true. The story was told with such conviction and detail, that they went along with every word. If they should raise a question about one of his tall tales, Shackleton would always counter with a convincing answer. And even if they still didn't believe a word of it, it was all such good fun that they were happy to be immersed in his world.

This Monument story indicates that from a young age Shackleton dreamed of performing a great deed, becoming a hero in the process, feted far and wide. He would spend the rest of his life trying to achieve exactly this.

It also highlights Shackleton's rare gift of telling a story and making people believe in it, and him. This was to prove an invaluable trait in years to come. It would allow him to earn people's trust, to fund his expeditions, to persuade others to do as he wished when he asked for the seemingly impossible, and even to make a living. For now, it was just the Shackleton sisters who believed his stories, and tended to his every need, but in the future this gift would see him have the world at his fingertips.

In his early years, despite his many dreams, it seemed it was only his family who had high hopes for him. Born in Ireland in 1874, his mother, Henrietta, was so taken by her ever-smiling boy, who seemed to always have a twinkle in his blue eyes, that she feared he was too good to live. With his sisters constantly running around after him, seemingly worshipping his every move, the young Shackleton certainly ruled the roost. This was a trait that endured his whole life, as his sister Kathleen later recalled: '"Come all my wives," he would shout when he entered the house after a voyage. He would lie down and call out: "You must entertain me. Zuleika, you may fan me. Fatima, tickle my toes. Come, oh favoured one and scratch my back." Of course we all loved it.'

In spite of this outpouring of love, the fortunes of the Shackleton family were on shaky ground. Since 1872, his father, Henry Shackleton, had worked as a farmer in Kilkea, County Kildare, just 30 miles from Dublin. However, by 1880 things were beginning to look bleak. The Americans, with a huge surplus of wheat from their prairies, had built a spiderweb of new roads to transport their grain to ports, where it could be exported at a minimal cost. Faced with such competition, agricultural depression soon followed in Europe. Henry saw that the writing was on the wall. Finally selling his farm in 1880, he moved his brood to Dublin, where he took up medicine at Trinity College Dublin.

Yet with Irish nationalism boiling over, and trouble brewing, as soon as Henry had completed his studies, he relocated the family to England in 1884, where he set up a practice in Croydon. After six months, they left Croydon and moved to Sydenham where Henry built up his business. Shackleton was now aged ten and quite used to being the centre of attention. That would quickly change. Upon attending Fir Lodge Preparatory School in Dulwich, he found himself outcast, teased for his Irish roots and slight brogue. Nicknamed 'Mick', Shackleton usually responded to such taunts with his fists. One classmate recalled, 'If there was a scrap he was usually in it.'

Although Shackleton eventually lost his accent and began to speak in a more southern tone, he would forever have to put up with the nickname Mick. However, he soon happily adopted it as his own, in later years even signing letters Mickey, giving the bullies little ammunition with which to taunt him. As we shall soon see, Shackleton's ability to bob and weave through the pitfalls of life was one of his many talents.

At thirteen years of age, Shackleton attended the public school, Dulwich College, just a short walk from his home. Once more, he initially found himself the outcast, liable to join in with any scrap, now earning the nickname 'the Fighting Shackleton'. It seems he disliked team games, was no sports enthusiast and was lazy in class. School reports included the comments 'wants waking up', 'is rather listless', 'often sinks into idleness' and 'must remember the importance of accuracy'.

The only thing that truly interested Shackleton was literature. At home, his father encouraged his children to read poetry, with Shackleton becoming an admirer of Tennyson and able to quote verse after verse. He also loved reading stories, particularly tales of derring-do set in the far-flung realms of the British empire. A favourite was *Boy's Own* magazine, which he bought every Saturday for a penny. He also devoured books by Rider Haggard and Jules Verne, especially the adventures of Captain Nemo in *Twenty Thousand Leagues under the Sea*.

While Shackleton readily consumed fictional tales of heroic adventures, neither did he have to look too far for the real thing. At the end of the eighteenth century the British empire was the largest in history, covering a fifth of the Earth's landmass, with one in four people on Earth – over 400 million – classed as British subjects. In 1887, when Shackleton was thirteen, Britain's frenetic celebrations of Queen Victoria's Golden Jubilee saw patriotism and pride in the empire at their zenith. Any explorer who could bravely defy the odds and conquer new lands for Queen and country was to be exalted far and wide.

The likes of Sir Henry Morton Stanley, who made his name exploring remote and hazardous regions, made front-page news, as did the Indiana Jones-like Colonel Percy Fawcett, who had earned fame due to his search for a fabled lost city deep in the Brazilian jungle, where he had subsequently disappeared. A key factor that ensured such explorers became the most famous stars of their time was the passing of the Forster Education Act in 1870. This had made education compulsory for children aged between five and twelve, and enabled, for the first time, many of the working class to read. With more people now reading than ever before, they delighted in stories focusing on the brave and daring deeds of the empire's explorers and conquerors.

As Shackleton read about these explorers, he must have noticed how adored they were by the British public and establishment alike. For a young boy struggling to fit in at school this seemed the answer to his prayers.

Having honed his gift for storytelling with his sisters, he now proceeded to tell fantastical stories to his classmates, whether his own, or read directly from the pages of

Boy's Own. Such was his storytelling knack, he soon earned a gang of followers who were happy to play truant in the local woods so that Shackleton could regale them with his tales.

When Shackleton told his gang a particularly thrilling story set at sea, such was the fervour he created that the boys immediately set off for London, where they prowled the docks, hoping to get jobs as cabin boys. To their frustration, and humiliation, they were instead sent packing.

Yet this trip to the docks lit a spark in Shackleton. Upon seeing the boats, all setting off for exotic locations, he realized that a life at sea held the key to his dreams of adventure. Most of his friends only really day-dreamed of such things, but he truly meant it. One of his many sisters later commented, 'He had no particular hobbies as a boy, but anything to do with the sea was his special attraction.'

Another event around this time might also have inspired him to expand his horizons. Soon after the Shackleton family's move to Sydenham his mother became sick, and spent the following forty years more or less confined to her bedroom. Seeing his beloved mother trapped in such a manner, through no fault of her own, perhaps made Shackleton realize that if he wanted to see the world, there wasn't a second to lose. Perhaps he might one day meet his mother's fate, and the world which now seemed so exciting and endless would forever be confined to the four walls of a bedroom.

Shackleton had also learned from his father's example that doing what might seem sensible did not always guarantee a happy life. His upbringing may have been relatively comfortable, but Henry Shackleton certainly sailed close to the wind a few times during his career as a farmer and then as a doctor. Shackleton may well have thought, why submit to a 'sensible' profession if it was just as precarious as an exciting one?

So, when it came time to leave school, Shackleton announced he was ready to explore the world on the ocean waves. Henry Shackleton was not amused. The family could not afford the Royal Navy, and he had hoped his son would follow in his footsteps and be a doctor, but he also knew how determined the stubborn Ernest could be. Grudgingly, he gave in, but at the same time, he had a plan of his own. He knew that if his son's ocean-going apprenticeship was sufficiently unpleasant, he would willingly switch to medicine on his return. Henry Shackleton therefore looked to sign his son on to one of the most testing apprenticeships the sea had to offer.

Remembering that his cousin Revd G. W. Woosnam was the superintendent of the Mersey Mission to Seamen in Liverpool, he asked him to utilize his contacts at the docks. A berth for Shackleton was subsequently found on board the *Hoghton Tower*, a cargo-carrying three-masted sailing ship. This would see him travel across some of the most treacherous seas in the world, alongside as rough and rugged a crew as you could imagine. And all for just a shilling a month. 'No fool,' Shackleton later wrote, 'my father thought to cure me of my predilection for the sea by letting me go in the most primitive manner possible as a "boy" on board a sailing ship.' Shackleton didn't care. At long last he was on his way, and his adventures could begin.

Shackleton's last three months at school saw an immediate change in his attitude to his studies, especially in mathematics, since it provided the basics of navigation. His maths teacher subsequently reported, 'He has given much satisfaction in every way. There has been a marked improvement both in his work and in his behaviour.'

Still, this didn't persuade Shackleton to stick around. He left school as soon as he could, finally saying his goodbyes at the tender age of sixteen in April 1890 and making his way to Liverpool to begin the first of his life's many great adventures. However, while he might have been bubbling with excitement at the prospect of freedom and adventure, it would prove to be an experience that would almost cost him his life.

2

Shackleton's introduction to life at sea was just as traumatic as his father had planned. In the past, he might have admired the boats on the Thames and envied those on board; now the reality was very different.

Carrying 2,000 tons of cargo, the *Hoghton Tower* embarked for Cape Horn, Chile, a 20,000-mile journey, in the middle of the southern hemisphere winter. A byword for danger at the best of times, this voyage put even the most well-built vessels to the test. The *Hoghton Tower* might once have been described as 'a magnificent specimen of iron shipbuilding', boasting accommodation for sixteen first-class passengers, but her days of luxury were now far behind her. In the age of steam, the three-masted ship was now very much a relic, picking up only those jobs that steamers avoided like the plague.

Facing howling winds, crashing seas, monstrous waves and even icebergs, the ageing ship struggled to stay afloat, losing two lifeboats off the Horn, with several of the crew injured. Shackleton spent most of the storm with his head over the side, vomiting, his sea legs apparently deserting him. But seasickness was the least of his worries.

Ever since they set sail, on 30 April 1890, Shackleton, the boy from a middle-class, suburban background, had found things tough. Sharing cramped quarters with the foul-mouthed, drunken crew, an appalled Shackleton called them 'lower than beasts'. A mixture of vagabonds, chancers and grizzled sailors, each had a story to tell, as Shackleton later revealed in a letter to a friend: 'There is an American who had to flee his country for killing a coloured man; another who was foreman in a large timber works . . . another owner of a large cattle ranch . . . Only the other day I saw a man stab another with a knife in the thigh right up to the handle.' For a young boy expecting his first voyage to be like something out of *Boy's Own* magazine, this was indeed a shock to the senses.

In those early days, Shackleton must have wondered if he was cut out for life at sea after all. Perhaps his father had been right all along? In another letter to an old schoolfriend he admitted it was 'pretty hard work and dirty work too . . . I can tell you it is not all honey at sea.'

Matters only got worse when the crew realized that the 'boy' disapproved of their behaviour. Just as had been the case at school, he was taunted and tagged a weirdo. Some no doubt felt that such ritual bullying was necessary to harden up the fresh-faced boy, while others merely enjoyed taking out life's frustrations on someone apparently vulnerable.

Finding little in common with the surly crew, Shackleton was often to be found alone, immersing himself in his books about adventure, exploration and empire-building, along with endlessly memorizing favourite poems, not to mention passages from the Bible. 'I learnt more of literature in a year at sea than I did in half a dozen years at school,' he later recalled. Indeed, whenever someone might have been looking for him, colleagues remember the familiar answer: 'Old Shack's busy with his books.'

Despite his quiet and withdrawn nature, the boy once known to his classmates as the Fighting Shackleton was certainly no pushover. When one crew member kicked him in the leg, Shackleton fell to the deck and bit deep into his tormentor's shin. It was a move that earned the grudging respect of the crew, while ensuring that anyone would think twice before getting physical with the boy, who clearly had hot blood coursing through his veins.

As the ship pushed out to Chile and Shackleton acclimatized to this new way of life, he slowly ingratiated himself with many of the crew, just as he had done with his classmates at school. Moreover, he found that his deep knowledge of literature, poetry and religion along with his ability to tell a good story were valuable commodities at sea. Before long, the skipper noticed a handful of his more hardened crew crowding around the young boy as he reeled off one story after another, often leaving them in fits of laughter. Perhaps even more surprising was the fact that some even began to request verses from the Bible. Equally unexpected was that the young, seemingly virtuous boy was now telling his stories with a cigarette dangling from his lip and also letting the occasional profanity fly. Growing ever more comfortable and confident, Shackleton moved through the different hierarchies of the ship with apparent ease, telling his tales to crew and captain alike. Some might have been suspicious that he was perhaps seeking to further himself, but it was clear that he did this for no more reward than the fact that he enjoyed company, whomever it might be. As one sailor later noted of him, he was 'several types bound in one volume'.

While his storytelling ability softened many of the crew's attitudes towards him, the quality they truly appreciated was his willingness to get stuck into the hard graft. There was no glamour in many of the tasks he had to do, but Shackleton proved to be eager, whether it be scrubbing the decks, tying knots, moving heavy cargo in and out of storage holds, or even climbing the 150-foot-high mast in the eye of a storm. Never uttering a quibble or a complaint, Shackleton just got on with the job at hand, keen to show his worth. Such was his impact that when the battered ship and crew arrived in Valparaíso after fifteen weeks at sea, the skipper, Captain Partridge, invited him to dinner with the local consul as a reward.

When the *Hoghton Tower* arrived back in Liverpool, almost a year after she had left, Shackleton said it was 'one of the stiffest apprenticeships' a boy could ever experience. However, was it enough to put him off a career at sea, as his father had planned?

Upon returning home to Sydenham, where he regaled his sisters with fantastic stories of his adventures and enjoyed home cooking and a cosy bed, his father casually asked his son about his future plans. Without missing a beat, Shackleton answered that he remained undeterred in his ambition. He wanted to continue at sea and rise through the ranks as quickly as possible. This time, Henry Shackleton did not try to stop him. In his year away, his son had grown to be a man, so if this was to be his choice, then so be it. He clearly knew what he was getting himself into.

Captain Partridge was also keen to have him back. While he said that Shackleton was 'the most pig-headed, obstinate boy I have ever come across', he also told Revd Woosnam that there was 'no real fault to find with him'. Shackleton soon signed his indentures for four more years at sea and was all set to return to the *Hoghton Tower*, albeit now with Captain Robert Robinson at the helm.

Setting sail from Cardiff in June 1891, this journey to Chile was to be even harder than the first. Again, the weather around Cape Horn was formidable, with one man washed overboard and lost, while eight others were badly injured in accidents. Shackleton was also forced to lie for days in his wet bunk suffering from agonizing back pain brought on by a month spent in soaked clothing. On landing in Chile, he was then attacked by an acute bout of dysentery, making the journey even more unpleasant. But Shackleton's biggest gripe concerned Captain Robinson, who was a far harder taskmaster than Captain Partridge had been.

A little under a year later, Shackleton made it clear to his family that he hated the *Hoghton*, the skipper and most of the crew. Nonetheless, he knew that his chances of advancement would be dented unless he completed a third trip. At least, this time, the initial destination was India, not Chile. However, it was during this trip that Shackleton nearly lost his life.

When a crew member died suddenly, some of the more superstitious crew told Shackleton that it was a sign that an evil omen was aboard. Shackleton might have scoffed at such tales, but soon after a storm struck without warning just south of the Cape of Good Hope. Such was its ferocity there was no time to take in the sails. At the wheel, Shackleton was desperately trying to keep the ship upright and on course when a huge wave keeled her over and flooded the decks. As Shackleton remembered, 'Nature seemed to be pouring out the vials of her wrath.'

Gasping for breath, choking on salty water, he heard a cracking sound from above. Moments later the mast came crashing down, slamming into the area where Shackleton had been standing shortly before. Counting his blessings, he later wrote, 'It was a miracle that I was not killed.'

After surviving further storms in two years at sea, going from India, to Australia, to Chile, he eventually completed his apprenticeship in July 1894. It had been four years since he had set forth on his first journey and in that time, he had truly grown to be a man. Blossoming from a fresh-faced, uneasy youth, he was now a strapping twenty-year-old with the confidence of having survived all that life at sea had thrown at him. Now with added muscle, a strong jaw and a sense of purpose, his many months spent with an assortment of characters had also provided many more yarns to his arsenal, now told with ever more elucidation and sparkle, in a rich tone that could only be attributable to a man of the sea.

Despite having found the last journey aboard the *Hoghton* particularly unforgiving, Shackleton certainly had no qualms about continuing in his chosen profession. No doubt the lack of any alternative career might have also played a large part in this decision, but nonetheless he soon passed the Board of Trade examination as second mate and subsequently took a posting as third mate on the *Monmouthshire*, a tramp streamer with the Shire Line.

This was a considerable step up from the outdated *Hoghton Tower*. In particular, on her regular journeys to China and Japan, she offered him the luxury of his own cabin, where he could read in peace. In 1896, he passed his exams for first mate and was then upgraded again, this time to serve on the even more luxurious *Flintshire*.

By the age of twenty-three, Shackleton had already seen much of the world and had a decent profession, even if it wasn't one of which his father particularly approved. It seemed that a lifetime of adventures on the sea awaited him. But soon his eyes turned to another stormy adventure altogether, that of love and romance.

3

In July 1897 Ernest Shackleton met the woman who would inspire him to greatness. As his father tended to his blossoming roses in the Sydenham sunshine, Shackleton, now on home leave, barely paid them any notice. The young sailor's attention was instead firmly focused on one of his sister's guests.

The woman in question was strikingly attractive; of that there was no doubt. A slender brunette with piercing blue eyes, her easy smile and elegant stature certainly beguiled the young Irishman. But after being introduced by his sisters and learning that the lady in question was called Emily Dorman, he soon found that there was more to her than met the eye.

The daughter of a wealthy and successful solicitor, Emily was well educated and intelligent, able to flit between flirtation and keen debate in a blink of an eye. She had already captured many a man's attention, turning down as many as sixteen proposals of marriage in the process. Rarer still, in an age when people married young, was the fact that she remained single in her late twenties. It was clear that she would not accept a partner for the sake of it. She was far too independent of mind and free-spirited to accept such a fate. Only a man who was capable of sweeping her off her feet would come under consideration.

At six years her junior, Shackleton eagerly took up the challenge. It certainly seems that he was genuinely smitten with Emily, but perhaps his spirit of adventure also played a role. Like so many of his future expeditions, Emily was an unconquered, uncharted challenge. Many men had tried to gain her attention and had failed. She would clearly not accept second best. So, if Shackleton were to catch her eye, and maybe even earn her hand in marriage, then he would have succeeded where so many others had failed.

Yet despite his best efforts to be at his most charming, Emily was clearly not struck by Cupid's arrow. Greeting him coolly, she barely paid him any more attention than was necessary for the sake of politeness. And why should she? On paper, Shackleton was not only younger than she was, he was also just a modest ship's officer. At the tender age of twenty-three, he might have passed his Master's Certificate, which, in theory, enabled him to take command of any Merchant Navy vessel anywhere in the world, but against competing suitors from high society this held very little sway.

Undeterred by Emily's cool manner, and with his pursuit interrupted by journeys at sea, Shackleton never lost sight of the fact that this was the woman he wanted to marry. Whenever he returned from one of his many voyages, Shackleton continued to pursue her relentlessly, and while she was initially suspicious that she might just be the latest temporary crush of a restless seaman, she had to admit that the barrel-chested young man did have an intriguing hint of mischief and adventure about him.

After months of Shackleton pulling out all the stops, Emily eventually agreed to accompany him to the British Museum. To her surprise, she found that beneath the blarney was a sensitive young man with a love of the arts, particularly poetry. Shackleton was of course well practised in spinning stories and reciting poetry to get even the hardest men on side, and his magnetic way with words soon captured Emily's attention. While Shackleton quoted long verses of Tennyson, verbatim, Emily shot back verses from her favourite poet, Robert Browning, whom she had studied on a university course. However, having written a paper on Browning's poem 'Paracelsus', she might have already recognized warning signs about any future relationship with Shackleton, as it told of a hero who put aside love in order to seek knowledge.

With their love of poetry serving as a common ground, Emily gave Shackleton a biography of Browning for him to read at sea. Soon, he also became an avid admirer of the poet, later writing, 'I tell you what I find in Browning is a consistent, a spontaneous optimism. No poet ever met the riddle of the universe with a more radiant answer. He knows what the universe expects of man – courage, endurance, faith – faith in the goodness of existence.' In a letter to Emily, he even quoted a line from Browning's 'The Statue and the Bust': 'Let a man contend to the uttermost for his life's set prize.' This left her in no doubt that she was the woman the young sailor wanted above all else.

However, just as Emily seemed to grow close, she would inexplicably pull away, playing havoc with Shackleton's emotions. As he wrote in another letter to her, sent while at sea, the future is 'so uncertain that I dare hardly shape a hope'. Increasingly desperate to earn her affections, and nervous that there might be competition, Shackleton was beside himself when, around Christmas 1898, the *Flintshire* prepared to set off on another long journey. Faced with the prospect of yet more months away, Shackleton was desolate as the ship left port, fearing that he might have lost his chance. Fate, however, dealt Shackleton an unexpected reprieve.

On Boxing Day 1898, the *Flintshire* ran aground just off the Yorkshire coast and required repairs. Sensing an opportunity, Shackleton left the ship and made straight for the

Dormans' house. Eager to take his chance, Shackleton threw caution to the wind and declared his love for Emily in the privacy of the billiard room. For the first time, Emily could tell that the young sailor meant what he said. Suddenly, a future together no longer seemed so ludicrous.

However, there remained another obstacle before Shackleton and Emily could be married: her father. It wasn't that Charles Dorman disliked Shackleton. On the contrary, he found him polite and charming company. It was his chosen profession that was the problem. A life at sea was far removed from that of a doctor or solicitor. It was certainly not as well regarded, or well paid, and it also meant being away for months at a time. No matter his affection for Shackleton, this was hardly the life he had in mind for his daughter, who had seen off far more eligible bachelors in the past.

Despite these misgivings, Shackleton was not cast out into the cold. When not at sea, he spent many a weekend at the Dormans' farm in East Sussex, and he was also invited to dinner parties, at which he was ever eager to impress his prospective father-in-law. Emily's niece remembered: 'He was so nice to the maids when they waited at the table, they almost dropped the things because he made them laugh so.'

One particular story, however, highlights Shackleton's own concerns that he might not measure up. While travelling by train to the Dormans, he struck up a conversation with an antique dealer who he found was attending an auction at the Dorman home. When the dealer asked, 'What are you hoping to get out of the old man?', Shackleton replied, 'His daughter, I hope.'

I can certainly sympathize with Shackleton's predicament. As a young man, I had set my sights on my late wife, Ginny, whom I had known since we were both children. Her father was understandably very protective of his beautiful daughter, particularly from the likes of me.

Like Shackleton, I seemed to offer very little. While I went to Eton, I did not excel academically and was all set for a career in the armed forces. Ginny, on the other hand, went to an all-girls boarding school and could have had any man she wanted. After a series of misdemeanours, which culminated in me attempting to blow up the film set of *Dr Dolittle* in a misjudged environmental protest, Ginny's father banned her from seeing me, telling my mother that I was 'mad, bad and dangerous to know'. Still, this didn't deter me. No matter what, I was determined that we would be together and, while it took many years, by the time we were married, her father had come to accept me in some measure at least, realizing that, like Shackleton, there was no putting me off. It also helped that Ginny returned my love.

While I was somewhat of a bull in a china shop, Shackleton went about things far more smoothly. If he was to be worthy of Emily and gain her father's approval, he knew he had to do better than a life at sea. During a conversation with James Dunsmore, the *Flintshire*'s engineer, Shackleton said, 'You see, old man, as long as I remain with this company I will never be more than a skipper. But I think I can do something better. In fact, I would like to make a name for myself – and for her.'

With this in mind, and soon after pouring his heart out to Emily in the Dorman billiard room, Shackleton resigned from Shire Line and looked for a more impressive berth. He quickly found it with Castle Line, a prestigious shipping company, whose 5,000-ton passenger liner *Tantallon Castle* was the cream of the crop. Steeped in luxury, its clientele was often made up of a Who's Who of British society.

Serving as fourth officer, Shackleton embarked on three

trips to South Africa, finding the surroundings on the *Tantallon* far more in keeping with his lofty ambitions. And, while his new role proved that he might yet be worthy of Emily, he also hoped it might open many doors.

Armed with a natural confidence, he was now well aware that he had a rare gift to make people like him, and he intended to use it to his advantage. While on his early voyages at sea he had used his gifts to fit in with the *Hoghton Tower*'s crew, here he made it his mission to ingratiate himself with high society.

Perfecting a chameleon-like ability to transform his character and tales to suit any particular audience, he was able to move between the roguish Irish 'mickey', the hot-blooded sailor with tales to burn and the middle-class gentleman with a vast knowledge of the world, charming all who crossed his path. Veering from quoting lines of poetry to sharing more bawdy tales of his seagoing adventures, Shackleton had the unique ability to make a new acquaintance feel at ease. His piercing gaze and infectious enthusiasm saw one of the Castle Line skippers say of him: 'His eyes were bright and his glances quick . . . On a subject that absorbed his interest, his voice changed to a deep, vibrant tone, his eyes shone, and he showed that determined, self-reliant, fearless and dominant personality which, later, was to make him a leader men would obey and follow unhesitatingly.'

Prosperous businessmen were not always known for being easily impressed, but Shackleton's nature soon disarmed them. He did it all with such ease, and confidence, that no one felt they were being used for any purpose other than friendship. More often than not, that was just the case, as Shackleton welcomed the opportunity for company, it didn't matter what class they happened to be from. Rather than be seen as just a member of staff, Shackleton's presence was actively encouraged at most tables, with most recognizing his ability to lift any mood. One such admirer was the steel magnate Gerald Lysaght, who in time would become one of Shackleton's great supporters.

Such luxury, and company, would sadly not last long. In October 1899, war broke out between Britain and the Afrikaner-speaking settlers in South Africa. While Shackleton was promoted to third officer on the 3,500-ton *Tintagel Castle*, high-society passengers were now replaced by thousands of troops who required passage to the Cape to fend off the Boer guerrillas. Among those who answered the call was Shackleton's younger brother, Frank.

On 14 December 1899, the *Tintagel Castle* set off from Southampton for the Cape, carrying a cargo of 1,200 troops, with high hopes that the war would be over in no time at all. Still, Shackleton charmed and entertained everyone he met. With thousands of restless young men looking for amusement, his ability to tell a story was always in high demand, while he also organized sporting activities and concerts. In his spare time, as always, he read a great deal, with a friend describing Shackleton's cabin as 'having a bookcase with the signs of a well-read owner for, in it, I saw Shakespeare, Longfellow, Darwin and Dickens'.

The war in South Africa did not, however, go as hoped. Over 3,000 British soldiers were killed, wounded or captured in the first week of conflict. The *Tintagel Castle*, along with Shackleton, was therefore required to do a quick turn-around and swiftly return to the Cape to supply ever more soldiers for the battlefield.

Always one to spot an opportunity amidst the apparent gloom, Shackleton soon recognized a way in which he could make his name and impress Emily and her father. Along with the ship's doctor, William McLean, and others, he produced a small book about their trips to South Africa, titled, *OHMS or How 1200 Soldiers Went to Table Bay*. With excerpts of some of his favourite poems also included, two thousand subscribers eagerly snapped up copies at 2 shillings and 6 pence. So proud was he of this new venture that he sent a specially bound copy to Queen Victoria, although his greatest thrill of all was being able to present one to Emily, inscribed 'E to E July 1900 The First Fruits'.

Despite the book's success, and his sterling work with Castle Line, the fact that Shackleton inscribed Emily's copy of the book with 'The First Fruits' hints that he felt he was still unworthy of her. Indeed, during his long voyages at sea he continued to believe that there were other suitors nipping at her heels, which caused him great distress. In letters to her he wrote:

> I suppose it is [a] man's way to want a woman altogether to himself . . . I said it in the old days 'Love me only a little, just a little' and now it seems as I grow older I am saying 'Love me altogether and only me' . . . When like today you spoke of him: something catches at my heart and I feel lost, out in the cold . . . Why did I not know you first? Why did you not tremble to my touch first of all the men in the world? Of all tales of love and sorrow I feel ours stands out for there was no hope in the beginning and there is none now.

Thankfully, an opportunity soon presented itself that would allow Shackleton to live out his dreams and impress both Emily and her father. It would, however, involve a trip to the other side of the world, where no man had ever before set foot.

PART TWO

‘A soul whipped on by the wanderfire’

4

In the summer of 1900, an advertisement caught Shackleton's eye. Under the command of the Royal Navy's Lieutenant Robert Scott, the Royal Geographical Society (RGS) was financing an expedition to the Antarctic. The advert was therefore seeking to recruit suitable crew members. Shackleton had neither the exploration nor scientific experience, but this was just the sort of opportunity he had spent his whole life looking for.

Straight from the pages of *Boy's Own*, the chance to explore a largely uncharted continent offered adventure, fame and fortune. Moreover, with the British empire thirsting for more conquests in the face of competition from the likes of America and Germany, any man who was to conquer an unknown continent such as Antarctica and claim it for Queen and country was sure to be celebrated as a hero for decades to come.

It was all of this, rather than the chance to actually explore the Antarctic, that truly drove Shackleton. Louis Bernacchi, the physicist and astronomer on the expedition, later remembered that Shackleton 'evinced no interest in any previous Antarctic expeditions' and was 'hungry for adventure and fame'. One of Shackleton's shipmates on the Castle Line continued on the same theme: 'He was attracted by the opportunity of breaking away from the monotony of method and routine – from an existence which might eventually strangle his individuality. He saw himself so slowly progressing to the command of a liner that his spirit rebelled at the thought of the best years of his life and virility passing away in weary waiting.' Shackleton's first biographer, Hugh Robert Mill, later added that Antarctica was 'an opportunity and nothing more'.

One of the other primary attractions of the expedition was of course the opportunity for Shackleton to prove himself worthy of Emily. As Shackleton's daughter, Cecily, was to later comment about her father's pursuit of her mother during this time, 'He was going to lay the world at her feet, and be worthy of this very lovely . . . woman, with brilliant blue eyes and wonderful smile . . . He wanted to pour something out round her feet and say, "There you are, you see, you've married a man who's making his own way in life and I've brought you back the goods."'

Brimming with excitement, Shackleton formally applied to join the expedition in September 1900. However, if he had taken the time to learn of the fate of many of the previous expeditions to the ice, he might well have had second thoughts.

The Antarctic Circle was first crossed by Captain James Cook in 1773, yet he never actually saw the continent of Antarctica itself. Due to hundreds of miles of pack ice, his ship, the *Resolution*, was blocked from venturing any further. While land was finally sighted in 1820, and the continent circumnavigated by Bellinghausen in 1820–21, it was not until the 1840s that two Royal Navy ships, *Erebus* and *Terror*, were able to break through the pack and explore more fully. This honour fell to James Clark Ross and Francis Crozier, who, as well as working out the approximate whereabouts of the South Magnetic Pole, were able to map geographical features for the first time, including the Ross Sea, McMurdo Sound and the Great Ice Barrier, known today as the Ross Ice Shelf. However, a disastrous Royal Navy expedition in the Arctic

North would shortly afterwards put a stop to any further British polar exploration for thirty years.

This mishap occurred in 1845, during Sir John Franklin's expedition to map the last unnavigated section of the North-west Passage in the Canadian Arctic. Working their way through the pack ice, the two ships *Erebus* and *Terror* suddenly became stuck. Rescue missions were frantically put together to search for the missing 129 men, at huge cost – eventually topping £700,000 (£40 million today) – but to no avail.

This cost, and subsequent scars, meant that it would not be until 1874 that a British boat would return to the south, when the *Challenger* expedition crossed the Antarctic Circle. Yet, even after this, so many questions remained unanswered. Was the continent a huge land mass, a chain of islands, or just a giant block of ice? So little was known that it might as well have been Mars. As the *Daily Express* later commented, 'its environment remains a kind of silence and mist and vague terrors.'

The oceanographer Sir John Murray soon grew frustrated at the lack of progress towards answers to these questions. Having served as a naturalist on the *Challenger* expedition, and with the disaster of the Franklin expedition fading from memory, he addressed the Royal Geographical Society in 1893, urging them to renew the exploration of the Antarctic. In the audience, listening intently, was Sir Clements Markham. President of the Royal Geographical Society (RGS), Markham was a former Royal Navy man who in 1850 had joined a rescue voyage to the Arctic with the aim of locating the ships lost in the Franklin expedition. Inspired by Murray's call to arms, he became obsessed with the notion that Britain should lead the way in Antarctic exploration.

Immediately setting to work on this National Antarctic Expedition, Markham mapped out a plan that would be executed by the Royal Navy. The first step would see a ship force its way through the pack ice to find a suitable anchorage for the crew to overwinter on board. In the summer, the Royal Navy crew would 'explore this Antarctic continent by land, to ascertain its physical features and, above all, to discover the character of its rocks and to find fossils to throw a light on its geological history'. There was as yet no mention of trying to reach the South Pole, but Markham's great hope was that a British man, from the Royal Navy no less, should be the first to claim this holy grail.

In 1895, Markham's project was approved at the Sixth International Geographical Congress in London. The congress exclaimed that it was the 'greatest piece of geographical exploration still to be undertaken' and that it would advance knowledge in 'almost every branch' of science. Others were not quite as enthusiastic.

When it became apparent that funds of over £100,000 (£6 million today) would need to be raised, Markham found that the Prime Minister, the Admiralty and even the Royal Society were reluctant to become involved. Memories of the tragic fate, and cost, of the Franklin expedition lingered and Markham was told that huge sums of money were required to modernize the Navy fleet, especially with Germany's fleet undergoing great improvements. In short, there was no money available to finance such a grand folly.

Infighting between politicians, the Navy, the RGS and the Royal Society didn't help matters either. All disagreed as to the expedition's specific aims, some believing that the expedition should concentrate on the glory of claiming uncharted land for Queen and country while others felt that a scientific expedition would be of far greater value. Markham could do nothing but watch as other countries and explorers sought to beat him, and Britain, to the punch.

In January 1895, the Norwegian Carsten Borchgrevink claimed to be the first man to have set foot on the Antarctic continent. Three years later, a Belgian expedition set out to cross the Antarctic Circle and became the first party to overwinter on the continent, albeit unwittingly, as their ship, the *Belgica*, became trapped in the ice.

However, for Markham, the real twist of the knife came in 1898, when a rival British expedition set off for the Antarctic. The *Southern Cross* expedition was privately financed by the British magazine publisher Sir George Newnes, and it was an enormous success, being the first to overwinter on the Antarctic mainland itself, as well as the first to visit the Great Ice Barrier since Sir James Clark Ross's ground-breaking expedition of 1839–43. In doing so, it also pioneered the use of dogs and sledges in Antarctic travel. To many, this had torpedoed Markham's plans for his own expedition, which was now known as 'Discovery'.

The end of the century drew ever closer and Markham had still raised only £14,000, with no further donations in sight. The government and the Royal Society remained uninterested and neither had a private benefactor stepped forward. Meanwhile, the German government announced that it was going to finance the Gauss expedition to Antarctica. However, just when all seemed lost, the wealthy industrialist Llewellyn Wood Longstaff answered Markham's prayers and donated £25,000 (£1.5 million today).

Shortly afterwards, the Prince of Wales, the future King Edward VII, agreed to become patron of the expedition, and his son, the future George V, vice-patron. I know from experience just how crucial a royal patron can be. In 1972, Ginny and I set about trying to raise money for our Trans Globe Expedition, needing to raise around £30 million in donations and sponsorship, all while having very limited expedition experience. At the time, we had £210 in the bank, owned a second-hand minivan and a heavily mortgaged, semi-detached house near Hammersmith Bridge. The only income we had was my infrequent pay cheques from the Territorial Army. Our dream seemed fanciful at best, and while we certainly surprised ourselves, and others, in raising a large sum of money, it wasn't until Prince Charles gave us his royal seal of approval in 1978 that people really began to take us seriously. Indeed, when we faced a serious issue in finding someone to sponsor our fuel, which would cost £1 million, it was thanks to the Prince that we eventually struck a deal with Mobil Oil. Prince Charles was to become the patron of many of my subsequent expeditions, and this opened further doors. In addition to this, he was also a great support, even going so far as to advise me on picking my team. In short, the support of a Royal can go a long way, as Markham would now find.

Once Markham's project had been approved by the Prince of Wales, the government felt it could no longer be seen to be holding out and offered Markham £45,000 towards the cost, on the proviso that he raised the rest from private sources.

Yet while Markham's dream seemed to be fast becoming a reality, things almost came crashing down. Markham's primary aim was for a geographic approach, exploring the unknown continent and mapping it accordingly. The Royal Society, however, championed a more scientific agenda, studying rocks and minerals, among other things. Although Markham had so far been able to balance this dispute, the disagreement came to the fore when discussions about who to appoint to lead the expedition began.

The Society advocated Professor John Walter Gregory, an acclaimed geologist and explorer. In contrast, Markham put forward Robert Falcon Scott, a thirty-one-year-old torpedo-lieutenant in the Royal Navy. Markham had first noticed Scott in 1887, in a race between two Navy ships in the West Indies. He obviously made a big impression, as Markham said, 'I was much struck by his intelligence, information and the charm of his manner.' And yet Scott had no experience of ice and had never before commanded an expedition. Despite this, Markham was determined to have his way, digging his heels in with a rabid determination until he tired the society into submission.

Lieutenant Scott was duly appointed as leader, and turned his attention to appointing his crew, on which Shackleton hoped to serve. However, with Shackleton's application now in Scott's hand, he read through the particulars then threw it into the bin and, along with it, all of Shackleton's hopes and dreams.

5

Shackleton was distraught. He had hoped that his experience at sea might have been enough for him to earn a berth on the merchant ship *Discovery*. But having no expedition experience, and not being enlisted in the Royal Navy, had proven fatal to his chances. The door appeared to be firmly closed. If he was to live out his dreams, as well as prove himself to Emily and her father, he'd have to think again. However, he soon realized that all might not be lost just yet.

Upon learning that Llewellyn Longstaff had financed much of the expedition, he recalled that on one his trips to the Cape he had met Longstaff's son Cedric. Using this relationship as a way in, Shackleton arranged to meet with Cedric's father at his home in the London suburb of Wimbledon.

Shackleton's time mixing with high society had served him well. While he proved to be charming company, in Llewelyn Longstaff he also found a kindred spirit, a man who seemed to love the prospect of adventure and risk almost as much as he did. After an enjoyable afternoon in each other's company, the wealthy industrialist promised Shackleton that he would personally recommend his services to Markham.

There was a slight hitch, in that Markham had wanted to staff the entire crew of the *Discovery* with Royal Navy men under a regime of standard naval discipline. However, he was in no position to turn down Longstaff's request, particularly as the expedition was likely to need more money in the future. As such, he merely promised that he would see to it that Scott would look once more at Shackleton's application, and this is where Shackleton now met with a stroke of luck.

Swamped with the deluge of preparations, Scott asked Albert Armitage, his thirty-six-year-old second-in-command, to take care of it. A fellow merchant marine officer, Armitage was impressed by Shackleton's experience, and he also saw something that Scott had missed. The *Discovery*, the ship for the expedition, was to be one of the last traditional wooden three-masted ships built in Britain, as steel steamships powered by coal were now in vogue. Shackleton was one of the few applicants who had experience on such a ship, owing to his time on the *Hoghton Tower*.

Armitage was also aware that an expedition of this sort required a particular personality. There was no room for troublemakers or shirkers. Everyone would be required to fit in and pull their weight. With this in mind, Armitage sought out testimonials from Shackleton's employer at Castle Line. Word soon came back that he was a 'very good fellow' and 'more intelligent than the average officer'. This was all Armitage needed to hear in order to persuade Scott that Shackleton was worthy of a berth.

In early 1901, while returning to England on the *Carisbrooke Castle*, Shackleton learned that Scott had accepted his application. As a junior officer, his tasks would include managing provisions and building the ship's library. For this he would be paid the meagre sum of £250 a year (around £15,000 today). Despite the poor pay, he was thrilled; such a job was worth far more to Shackleton than money. While he told Emily this was all for her, she was far too wise to fall for it. 'He used to say he went on the *Discovery* to get out of the "ruck" for me!' she wrote, 'it was dear of him to say it because I cannot flatter myself that it was only for me – it was his own spirit, a soul whipped on by the wanderfire.'

As soon as Shackleton arrived at Scott's London offices, he sought to impress. Desperate to be an integral member of the expedition, and eager to learn, he spoke to anyone who crossed his path. Whether they were Royal or Merchant Navy, and whatever their rank, or whether they were one of the five scientists who would also be on the trip, it was all the same to him. Thanks to his garrulous personality and gift of the gab, he was equally happy holding forth in the local pub with the young sailors, sharing bawdy stories in between a puff of a cigarette and a pint, as he was mixing with the scientists, asking questions about them and their interests. James Dell, a young sailor on the *Discovery*, observed that Shackleton was 'both fore and aft', while Clarence Hare, a steward, recalled that he was a 'good mixer'.

Shackleton's charm was also to provide dividends for the expedition's finances. On making a tour of the *Discovery*, Elizabeth Dawson-Lambton, a wealthy spinster, was so taken by him that she donated £1,000 to the expedition. The money was subsequently spent on a hot-air balloon, which Sir Joseph Hooker, the last remaining survivor of the Ross and Crozier expedition, had recommended. He had told Scott that if the expedition should find it was unable to travel as far as it desired, the balloon would still be able to provide unparalleled coastal views of the unexplored continent.

However, not everyone was in thrall to Shackleton's charm. The physicist Louis Bernacchi, who was to study magnetism on the continent, was particularly scathing. 'His aptitude for satire, for bantering his companions, could be embarrassing and sometimes annoying,' he wrote.

Yet, as had been the case on the *Hoghton Tower*, even those who were immune to the smiling and jocular Irishman could not deny that he was a hard worker. He was frequently seen lugging boxes up and down the office stairs, stacking them high to save space and then weighing them, to ensure they remained below 56 lbs so they could be easily unloaded when in Antarctica. His commitment and enthusiasm for any job was infectious. Markham commented that he was a 'marvel of intelligent energy', while even the critical Bernacchi was moved to concede that he was 'full of flashing new ideas'. Scott, the man whom Shackleton wanted to impress more than any other, was also complimentary, commending him for always being 'brimful of enthusiasm and good fellowship'.

On the surface, Scott, the Royal Navy officer, and Shackleton of the Merchant Marine could not have been more different. Scott, who had been at sea since he was thirteen, was more insular in character and more rigid in his approach, being a stickler for Navy codes of discipline. Some found him hard to warm to and felt he could at times be stubborn. In contrast, Shackleton loved the company of people and was a tornado of energy, sometimes to his own detriment. Rules did not particularly concern him; he relied instead on an innate sense of fairness.

However, Scott and Shackleton did have something important in common. Neither of them was particularly keen on polar exploration for the sake of excitement or the novelty of the great unknown. Rather, both hoped that the expedition would greatly benefit their futures, with Scott looking to further his career in the Navy and Shackleton to make his mark and win the hand of Emily. That they had embarked on polar exploration was a matter of chance; they would have seized any opportunity to get on with equal enthusiasm. However, they also shared the same tragic flaw, 'a colossal ignorance of their chosen field', with neither man having been to the ice before.

There was no denying that Scott faced a tremendous challenge in setting up a two- to three-year expedition with no prior experience. He had commenced preparations only in June 1900 and, with the expedition due to depart in August 1901, he had just fourteen months to organize one of the most ambitious expeditions ever mounted. Moreover, Scott could rely only on educated guesswork, since nobody knew what obstacles, what types of surface, or even what altitude they would encounter between the planned landing site at McMurdo Sound and the South Pole.

In comparison, I was very fortunate. I had previous polar explorers to consult, such as one of Britain's top polar experts, Sir Vivian Fuchs, who recommended that I spend a lot of time in Greenland and Northern Canada training on the ice before heading to the Antarctic. After doing so, while I was not exactly a polar veteran, I had at least mastered the art of deep-snow ice-cap travel in temperatures of minus 15° Celsius, and of traversing sea ice in temperatures of minus 40°. I was also able to learn vital skills, such as how to measure distance travelled by following shots of the sun at noon with a theodolite, the same tool Scott and Shackleton used on their expedition decades earlier. (My first attempt at this in Greenland was wrong by 62 nautical miles. If I had tried to master this skill for the first time in the Antarctic, it could have been disastrous.) All of this meant I was well prepared and knew somewhat what to expect, despite never before having been down south.

Scott did not have this luxury, and the obvious course was to assume that the conditions of the snow and ice in the south would mirror those in the north. This would mean that travel methods developed by past Royal Navy voyages in the North-West Passage's frozen zones could be copied. As a result, Scott knew that the likes of William Parry, James Clark Ross and Leopold McClintock had adapted certain Inuit travel systems, clocking up a number of highly impressive man-haul sledge journeys in the process.

However, since then, as Scott noted, 'England has not maintained her reputation in the sledging world', and 'it is abroad therefore that the modern traveller must go for all that is latest and best in this respect'. In the race to the North Pole, the Americans Charles Hall, Elisha Kane, Frederick Schwatka and Robert Peary had all followed various Inuit or Lapp practices, including travel with skis, dogs and light sledges. Scandinavian explorers Fridtjof Nansen, Otto Sverdrup and Roald Amundsen, who were accustomed to snow and ice in their own backyard, had also experimented with each of these methods. By contrast, Scott's own man Armitage advocated the use of Siberian ponies, rather than dog or manpower, relying heavily on his own past experiences on the ice in Franz Josef Land. Ponies, he said, could stand the severest cold and drag heavier loads than either dogs or men.

With much to ponder, Scott duly met up with the famous Norwegian polar explorer Fridtjof Nansen, who in 1895 had set a furthest north record. Nansen revealed that while he had mostly used dogs to pull the sledges in flat tundra regions, they were pretty useless in rough, broken terrain, where manpower was best. And Antarctica might, for all anyone knew, prove to have very rough terrain.

Although Nansen still recommended taking dogs, Markham was less than taken with the idea. Declaring dogs 'worse than useless', he wrote, 'In my mind no journey made with dogs can approach the height of that high conception which is raised when a party of men go forth to face hardships, dangers and difficulties with their own unaided effort. Surely in this case, the conquest is more nobly won.' Ultimately, while Scott took some dogs to help on the ice, he also agreed with Markham that the vast majority of the work would be done by the men themselves, man-hauling their heavily laden sledges across the region.

In making his decision Scott had sought out advice from as many experts as possible and had wisely chosen to hedge his bets between dogs and man-hauling. No one can criticize him for making it and, in the circumstances, I think he did rather well. Dogs were championed by those in the know, and he therefore took their advice, while also being prepared to man-haul.

Indeed, years later, when I was planning my own expeditions, I also sought advice on this matter from one of the foremost polar experts of my time, Wally Herbert. Herbert had led the only team ever to have crossed the Arctic Ocean, doing so with the aid of dogs, but he also had help via air re-supply. Despite this, Herbert wrote, 'The partnership of man and dog is the safest form of surface travel . . . If a dog dies, he and his food supply are eaten by his team-mates and the team carries on . . . Their performance has to be seen to be believed.'

I also discussed the matter with Dr Geoffrey Hattersley-Smith, one of the few men to have worked with machines and dogs along the edge of the Arctic Ocean. He also stressed to me that his dog team could move a great deal faster over patches of rough ice than any type of snow machine. Dogs do not refuse to start in low temperatures, nor do they break down and waste weeks, not just days, of good travel weather as a result.

However, despite his endorsement, Herbert also made it very clear that I should not use dogs unless I was prepared to spend a year or two learning how to handle them. I didn't fancy a two-year delay to achieve this, and I was also well aware that I was living in an era where the media might not look favourably on any perceived cruelty to animals. Indeed, a Japanese polar expedition had recently flown 180 huskies in specially built cages, only for 105 to die, something that garnered much criticism. Having to weigh up the pros and cons, I eventually decided to use snow machines to cross the Antarctic for the Trans Globe Expedition. Later, however, I would rely entirely on man-hauling.

While many have chosen to criticize the concept of man-hauling over the years, in 1993 my sledging partner, Mike Stroud, and I became the first men ever to cross the entire Antarctic continent doing exactly this, without a dog in sight. And if anyone should be shouting that perhaps our sledges were lighter, due to modern technology, that was certainly not the case. Each of us man-hauled a sledge weighing 485 lbs, which was in excess of anything Scott or Shackleton pulled. I say this not to boast about my own achievements but purely to set the record straight. Scott and Markham were right to believe that it was possible to reach the pole by man-hauling alone, even if Scott did take dogs as a precaution.

As preparations gathered pace, Shackleton was sent to Scotland to learn how to detonate explosives, a vital skill should the *Discovery* become trapped in the ice, as the stricken *Belgica* had been. He was also assigned to learn how to operate the hot-air balloon which had been paid for by Miss Dawson-Lambton's donation.

With the departure date fast approaching, Scott sent all of his men to the dentist, to ensure against any dental issues on the journey. It sounds as if this was a wise move, owing to the recorded removal of 92 teeth and the filling of 102 holes, at a cost of £62.45 (£3,000). I once did something similar when, in order to avoid the danger of any team member coming down with appendicitis while we were in the Antarctic, I recommended we all have our appendixes removed in London, prior to departure. To the great relief of some, this suggestion was turned down as more people die from the side effects of a general anaesthetic than they do from appendix troubles.

Tooth trouble was, however, the least of Shackleton's worries. The departure date was just weeks away and he had started to wonder if he was doing the right thing. While he was going to Antarctica to prove his worth to Emily, he was concerned that in being away for so long she might forget him. Worse still, she might meet someone else in the void. But it was too late to pull out now. All Shackleton could do was spend every spare moment with her and hope that he left a lasting impression.

Sixty-six years later I faced the same concerns as Shackleton. I was due to serve as a 'contract' officer in the Sultan of Oman's army for two years, and I was worried that I might lose Ginny. A week before I departed, I put an engagement ring on her finger in the hope it would ward off other suitors in my absence, almost as if I were planting a Union Jack at the pole. It seemed to do the trick as, although there were a few bumps in the road, we were married upon my return to Britain.

Despite his concerns, on 31 July 1901, as the *Discovery* set sail down the Thames, Shackleton was on board, watching the thousands of well-wishers lining the route cheering them on. Proudly sporting his new naval uniform, supplied by Markham, he could only imagine the reception that would greet them on their return if they were to be successful. Among the crowd he spotted his younger sisters Helen, Kathleen and Gladys. Waving his handkerchief, he signalled in semaphore, 'Goodbye Helen, goodbye Kathleen, goodbye Gladys.' But he saved his most emotional goodbye for Emily, waving vigorously, not knowing if and when he would see her again and, if he did, whether her heart would still be set on him.

Before the *Discovery* pushed out into the ocean, she still had one more vital stop to make. Docking in the Solent, a visit from royalty awaited the excited crew. King Edward VII (only recently on the throne, following the death of Queen Victoria in January 1901) and Queen Alexandra were on the Isle of Wight to celebrate Cowes Week, an annual social event for royalty and high society. Arrangements were, however, made so that the royals could wish the expedition luck, as well as award Scott with the Victorian Order.

Also in attendance were the proud Longstaff and Markham. After almost a decade of false dawns, Markham's expedition had now been given the highest honour, the royal seal of approval. Yet without the deep pockets of Longstaff, the expedition would have remained nothing more than a dream, as even in those final weeks it flirted with financial disaster. Once more Longstaff had ridden to the rescue, adding another £5,000 donation to the kitty. Meanwhile, watching in the background was Shackleton, who could barely believe he was rubbing shoulders with royalty. It was a sign that all of his dreams were on the right track.

Finally, on the morning of 7 August, the *Discovery* left England. The *Daily Express* reported that she was sailing 'with the hearty good wishes of all who hope that the race for the South Pole will be won by the first naval Power of the World'. Yet just before they departed, Shackleton had taken drastic, albeit uncharacteristic, measures to win Emily's hand in marriage.

After meeting the royals, and just days after waving Emily goodbye on the banks of the Thames, he had written to her father requesting his permission to marry his daughter on his return. Writing this last-ditch letter was not only a telling sign of Shackleton's deep concern that he might lose Emily, it also betrayed his lack of confidence by not asking Charles

Dorman in person. He had also perhaps hoped to capitalize on the mountain of good publicity that had wished the *Discovery* well on her departure. Surely this would impress Dorman and prove that he was indeed going to prove himself worthy of his daughter?

In Shackleton's letter he was open and honest. He was well aware of Dorman's reservations about him, and his profession. Of most concern was, of course, money. A life as a simple sailor didn't offer a particularly glittering future and would certainly not keep Emily in the style to which she had grown accustomed as the daughter of a successful solicitor. Shackleton addressed these concerns head on, writing that he had joined the expedition 'to get on . . . so that when I come back or later when I have made money I might with your permission marry Emily if she still cares for me as I feel she does now'.

Now setting off to sea, Shackleton knew he would not be able to receive any response, good or bad, for a number of weeks, when the *Discovery* landed in New Zealand. Until then, he faced an agonizing wait, as well as plenty more trials and tribulations.

6

The first few days aboard the *Discovery* were far from plain sailing. In fact, things were so bad that the expedition itself was put into question.

In order to overwinter in the Antarctic, it was vital that the ship reach the pack ice by late November. Any later, and winter would set in, with the route to the frozen coastline blocked off. But as the boat made its way to the Canary Islands, time was already against them. Described as a 'sluggish sailer', the *Discovery* was far slower than had been hoped, and she also devoured far more coal than had been envisaged. The ship's sails were also of little use, having not been tested before setting off. Shackleton's verdict was thus, 'Too much sail aft and not enough forward', while Charles Royds, the first lieutenant, commented that 'one or two great mistakes' had been made in the ship's construction.

On rough seas, the boat was a liability. In the Indian Ocean, she was 'tossed about like a cork', with many of the crew reaching for the sick buckets. Of far greater concern than a seasick crew was just how much water she took on board, with over 20 tons needing to be pumped out of the ship at least once a day. This almost culminated in disaster when Shackleton checked the holds to find them swamped, with many of the stores ruined. Although he rushed to save as much food as possible, try as he might, a vast quantity had to be thrown overboard. Owing to the stench of dirty water and rotting food, a new floor had to be built to store the remaining provisions. Scott was at least thankful to

Shackleton for his quick thinking. If much more had been lost, then the consequences would not have borne thinking about.

In 1979, we suffered a similar challenge. On our way to the Antarctic, our ship was invaded by a foul smell. Closer inspection revealed that our ancient fridge had failed, spewing out a ton of frozen mackerel which had turned to liquid. Our wolfskin parkas were stored in the same fridge and were forever impregnated with the evil stench of rotting fish.

In those early days, although Shackleton worked diligently, he continued to mingle with the crew, who were somewhat divided, owing to being a mix of Royal Navy and merchant seamen and scientists. Cliques subsequently developed, particularly as Scott had split the groups into specific living quarters. Ratings slept on the mess decks in hammocks, warrant officers in their own quarters and officers in individual cabins. While everyone ate the same food, they did so separately, with Navy stewards serving the officers in their wardroom on a table laid with linen and napkins. Shackleton paid no notice to any of this. He was happy to go from group to group, entertaining them as he went, although he was particularly drawn to the naval ratings and the scientists, hoping to learn whatever he could from their experiences in their particular fields.

One particularly close friendship he struck was with Hugh Robert Mill. While only travelling as far as Madeira, the forty-year-old scientist, who also acted as librarian at the Royal Geographical Society, was on board to teach some of the crew how to collect meteorological and oceanographic data. Shackleton was supposed to learn how to test the density and salinity of water, something Mill later recalled was a challenge for him: 'He found the minute accuracy required rather irksome and was long in grasping the importance of writing down one reading of an instrument before making the next.'

Nevertheless, while they made unlikely friends, one a patient and reserved scientist, the other an all-action adventurer, they enjoyed each other's company. Shackleton was always easy to warm to, thanks to his bubbling enthusiasm and energy, and he found Mill, and his profession, especially stimulating, which led to conversations long into the night. Mill wrote of their time together, 'it was on the bridge of the *Discovery* from midnight on, as the ship was rolling southward that I discovered his individuality and recognised how he differed in turn of phrase and trend of mind from the other splendid fellows [on board]'.

When Mill left the ship at Madeira, Shackleton quickly made friends with Dr Edward Wilson. A Cambridge graduate, Wilson had qualified in medicine only a year previously and had been working at Cheltenham General Hospital before joining the expedition. Now serving as the junior doctor, the twenty-nine-year-old was closer in age to Shackleton than Mill. He was also devoutly religious, so much so that one shipmate described him as 'the personification of Christ on earth'. On the surface, as was the case with Mill, the Oxbridge man of religion and the school dropout didn't have a lot in common. But away from the bawdy jokes and stories Shackleton liked to share with some of the crew, he found he could have an intelligent and in-depth discussion with Wilson on just about any subject. Soon the pair could be found on deck, late into the night, waxing lyrical under the stars. 'He has taken me in his charge,' Wilson wrote in his diary, 'and puts me up to endless tips and does no end of things for me.'

In early October, the heavily laden and leaking *Discovery* sluggishly arrived in Cape Town, more than a week behind schedule. While this was to have consequences for the expedition, Shackleton had also been hoping to meet his brother Frank, who was serving in South Africa with the Royal Irish Fusiliers. However, due to the *Discovery*'s late arrival, the two brothers missed each other, with Frank having been sent home wounded.

Scott now used the stopover at Cape Town to revise his plans to make up for lost time. Melbourne was crossed off the route and the *Discovery* would instead head directly for New Zealand. Shackleton had, however, previously arranged to pick up further supplies, as well as three prefabricated wooden huts, in Melbourne. He therefore had to urgently arrange for it all to be shipped to the New Zealand port of Lyttelton. While Scott hoped the more direct route would put them back on track, the forty-six-day voyage to Lyttelton would truly put the *Discovery* to the test.

In ferocious seas, the decks, and below, were again swamped, with the ship springing a leak and its steering gear jamming. Then, two days west of New Zealand, Scott recorded 'as big a sea as one ever comes across', with the inclinometer recording a record roll of 56°. In 1979, en route to the Antarctic, we borrowed the *Discovery*'s inclinometer and hung it on the side wall of our ship's bridge. One great Antarctic wave forced us to keel over 47° both ways, as recorded by the inclinometer's brass arrow. This was tremendously frightening, so one can only imagine how the crew of the *Discovery* felt in a far more vulnerable wooden ship, in the face of a far larger wave. I know I certainly feared for my life.

Despite the testing conditions, Scott's gamble had been worth it. In late November 1901, the *Discovery*, and her seasick crew, arrived at Lyttelton, on the South Island, dead on schedule. Yet on arrival, Shackleton had but one thing on his mind: had Charles Dorman responded to his letter?

At 2 a.m., the impatient Shackleton could not wait a

moment longer. It had been almost four agonizing months since he had sent the letter to Charles Dorman, and he needed to know his answer, if indeed there was one. Heading ashore, he urgently woke the postmaster, and to his relief he found two letters waiting for him, one from Charles Dorman, the other from Emily.

Hurriedly ripping open Charles Dorman's letter, he read the words quickly, his heart almost catching in his mouth: 'if and whenever the time comes when you are in the pecuniary position you so long for and that if you and Emmy are still of the same mind my consent to your union will not be wanting'.

Shackleton shouted for joy. Charles Dorman had given his approval for him to marry his daughter upon his return. Such were his celebrations, he almost woke the whole village, but there was to be a sad sting in the tale. Upon reading Emily's letter, he learned that shortly after Charles Dorman had sent his letter he had passed away. While saddened by this news, Shackleton was also relieved that one of Dorman's last acts had been to approve the union. He knew that Emily would also have found great comfort in this.

Shackleton now had no time to waste in readying the *Discovery* for the trials of the Antarctic. However, having sprung a few leaks en route to New Zealand, the *Discovery* had to be taken to a dry dock for repairs, although this at least allowed Shackleton some time to solve a major conundrum. With three years' worth of provisions, the hold was already filled to absolute capacity. But 30 tons of prefab huts, 40 tons of coal and 1,500 gallons of cooker paraffin had just arrived from Melbourne and room had somehow to be made for it all on board. Ultimately, Shackleton had no choice but to lash every item on the decks as best he could, which made the ship sit dangerously low in the water, as well as making her top heavy. They would have to pray that they avoided any storms on their way to the Antarctic, or the ship might become overwhelmed and sink.

Just before departure, Shackleton also had to find space for four dozen sheep donated by a local farmer. The only place for them was next to the twenty-three huskies. While Shackleton erected a barrier between them with sacks of coal, Scott, noticing the vicious snarling and barking of the dogs, decided to take his pet terrier, Scamp, off the ship for the duration of the voyage. Unbeknown to Scott, the dogs were from three different canine strains, which led to more aggression than is normal between sledge dogs.

In contrast to the fate of Scamp, Ginny's terrier, Bothie, travelled everywhere with her. He even became the first dog ever to have lifted his leg at both poles (a record the *Guinness Book of Records* politely declined to accept).

While Shackleton was kept busy loading the holds, Scott was faced with his first true test of leadership. With the crew taking advantage of their last chance for drunken debauchery, some had disgraced themselves, with Chief Engineer Skelton noting 'a great deal of fighting and drunkenness'. Scott was in a quandary. As the crew was a mix of Royal Navy and merchant seamen and scientists, he was not officially allowed to enforce the Naval Discipline Act. He was therefore well aware that he was very much walking a tightrope and his authority could come under question at any moment. Yet enforcing discipline on such an expedition was crucial, especially since he knew that previous polar expeditions such as those of the Norwegian Borchgrevink and the Belgian de Gerlache had been riven with dissent and near-mutiny.

Although Skelton recommended that two of the troublemakers be thrown off the ship, Scott was reluctant to do so; he was, in any case, shorthanded. After considering his options he took the middle ground and gave the two men a severe dressing-down, demoting them and making it very clear that, while he was in charge, he could also be merciful. It was a masterclass in leadership, and one the crew respected, with Louis Bernacchi extolling Scott's 'sense of right and justice'.

Despite this, one of the two miscreants subsequently deserted. Scott was, however, in luck. At that time, Tom Crean, a twenty-four-year-old Irishman, was serving as an able seaman in New Zealand aboard the torpedo vessel *Ringarooma.* Upon hearing that a space had opened up on the expedition, he eagerly offered his services. Scott was only too happy to have another Royal Navy man on board. Indeed, this was to be a fortuitous appointment, as Crean would not only make a valuable addition, he would later become one of Shackleton's most trustworthy sledging colleagues.

On 21 December 1901, the *Discovery* was ready to set sail for the Antarctic. Eager to witness such historic beginnings, the public flocked from all over the country to cheer them on their way. Yet as a brass band signalled their farewell, disaster struck. Enjoying the celebrations from the top of the mainmast, Charles Bonner, a twenty-three-year-old naval rating from London's East End, lost his grip and fell to his death. This tragedy saw the departure delayed for a few days, as while the crew was in shock the unfortunate Bonner also had to be buried.

It was therefore a very different scene when the cumbersome *Discovery* finally left Lyttelton on Christmas Eve, 1901. This time, the crowds and celebrations were far more muted, as were some of the crew. For as they set off into a gloom of thick fog, it dawned on them that they were heading towards a virtually undiscovered continent, where any number of great dangers lay in store. What's more, they would be alone, thousands of miles away from civilization, and rescue.

7

The further the crew plunged into the unknown, the greater became their sense of trepidation. Everyone was aware that the leaking, top-heavy ship would be in real trouble if they should encounter any bad weather, in what Scott described as 'the stormiest ocean in the world'. For now at least, as they travelled due south, passing the snow-covered mountains of South Victoria, there were only clear blue skies on the horizon.

On 3 January 1902, with the *Discovery* chugging its way through the Ross Sea, Shackleton found himself, for once, lost for words. Before him, almost glowing on the horizon, was the Antarctic Circle, a vast expanse of ice believed to be around 300–1,500 miles in width, stretching for as far as the eye could see. Taking in this phenomenal scene, on which very few men had ever before set their eyes, Shackleton wrote that he experienced a 'lack of facility' in order to adequately describe what he could see. All that lay before him now promised fame, fortune and adventure, not to mention the chance to be truly worthy of Emily. The thought of being the first man to reach the pole must have also crossed his mind, and the prospect of discovering new lands, wildlife and minerals for King and country. It was indeed the land of milk and honey.

However, it was here that the *Discovery* faced her first Antarctic challenge. Scott had hoped to breach the pack ice around late November, when the summer season would have melted a way forward. Now, such a path was blocked by constantly moving blocks of ice as big as cathedrals which at one moment would shut down open water, at the next tantalizingly reveal it again. It was like entering the jaws of death, and there was no guarantee they would find a way through. At any moment, the route could be blocked off and the ship trapped, a fate which had befallen the *Belgica* in 1898, with the crew forced to overwinter on the ship, and driven almost mad in the process. In such an environment, the fate of the *Discovery* was very much in the lap of the gods.

Scott, and most of the crew, might have known very little about preparing for the Antarctic, but he was at least as prepared as he could be in this respect. Having observed previous expedition boats, particularly Nansen's *Fram*, he knew how to maximize the *Discovery*'s chances of not becoming enmeshed in the ice. The ship had therefore been designed with a well-rounded, bulbous hull made from thick wooden beams which meant it could rise up without being crushed by the extreme pressure caused by million-ton ice floes on the move. This helped her plough a path through the miles of moving ice, seemingly untroubled.

As they chugged forwards, Scott halted the ship so that ornithological specimens could be collected from passing floes. He also used the opportunity to stock up on fresh meat for the winter, with some of the crew, including Shackleton, taking to the ice with rifles. This must have been an incredible experience for Shackleton. Never having fired a gun before, he was now on an ice floe hunting down penguins. In doing so he also shot a monstrous leopard seal near a ton in weight, which was then sliced up to hoist on board. Wilson, the appointed butcher, recalled that it was 'an enormous beast with a mouth full of teeth and a head bigger than a Polar Bear'. In time, Scott noted, 'we came thoroughly to enjoy our seal steaks and to revel in the thought of seal liver or kidneys.'

While food preparation remained vital, with all the Lyttelton sheep slaughtered, skinned and frozen, drinking water was also a valuable commodity. Scott had his men axe chunks of ice from the floes which could then be melted on board. This also allowed Scott to save his precious coal supplies by avoiding the use of the *Discovery*'s steam-driven desalination plant.

On 5 January 1902, with the *Discovery* still surrounded by the ice, Scott decided to belatedly celebrate Christmas Day, with the original festivity having been postponed after Bonner's death. As part of the Christmas fun, scientists, officers and men were all issued with skis and given their first chance to learn how to use them. The surviving photographs of the *Discovery*'s crew learning how to ski are generally hilarious, with as many people dotted about the ice floes on their backs as on their feet. Shackleton fared little better than most, commenting, 'I think my share of falls was greater than the others.' Scott had felt that skiing was in no way vital to the expedition, but he would soon learn that in certain circumstances it could be very useful. Yet, for now, the skis were very much for fun, and after exerting themselves on the ice everyone was back on board for a hearty meal, a sing-song and an extra rum ration. Suffice to say, spirits were high, and further good news was to come the next day.

In just four days and nights, the *Discovery* had come to the end of the ice belt. This was a phenomenal achievement, certainly when compared to other expeditions. In the Ross and Crozier expedition in the 1840s it had taken *Erebus* and *Terror* forty-six long days to break through the ice belt.

Leaving the belt behind, the *Discovery* continued to forge south, looking for a suitable place to anchor for the winter. On 9 January she passed the volcanic slopes of Cape Adare, which Bernacchi immediately recognized, being one of the few men on board who had been to Antarctica before, having overwintered there as a member of the *Southern Cross* expedition.

Taking a brief detour to go ashore, a small group inspected the *Southern Cross* expedition's hut, where they hoped to find useful stores or equipment that might have been left behind. Alas, Wilson described the decaying hut as looking like 'the centre of a rubbish heap'. However, it did have some use.

In 1902, cable telegraph existed, but it did not yet operate in remote regions. Ships at sea therefore had no way of informing anyone of changes of plan made en route. Selected sites, such as this expedition hut, were thus used as locations for a message post. If a ship should run into any trouble, the notes would inform any subsequent rescue party of where they had been and where they were intending to go. This location was one of five pre-selected sites, and Scott duly left behind a sealed canister containing a brief summary of his intentions.

The *Discovery* set sail once more, with Shackleton settling into the crow's nest, hoping to locate the most suitable location to set up base for the winter. From such a high vantage point, he could not fail to be dazzled by the view. Surrounded by ice glistening in the winter sun, ever larger white-tipped mountaintops and towering glaciers passed by every minute, as did whales, who swam alongside the ship spouting water high into the air. It was a phenomenal sight, even for an experienced seaman like Shackleton, who assumed he had seen all the sea had to offer.

On 20 January, soon after passing 76 degrees south, a small inlet surrounded by mountains seemed to offer a promising location. Soon after, Shackleton joined Scott and a few others on shore to inspect the area.

Scott named the inlet Granite Harbour but after some consideration deemed it unsuitable for their winter anchorage, believing a base could be found further south. However, the trip was certainly not wasted by Shackleton, who became the first person ever to spot plant life in the region. He later wrote of the incident:

> I was with Koettlitz [the expedition botanist], and seeing some green stuff at the foot of a boulder I called him to have a look at it. He went down on his knees and then jumped up, crying out, 'Moss!! Moss!! I have found Moss!!!' I said, 'Go on! I found it.' He took it quite seriously, and said, 'Never mind, it's moss; I am so glad.' The poor fellow was so overjoyed that there were almost tears in his eyes.

Soon after, they set sail for McMurdo Sound, named after Lieutenant Archibald McMurdo, who had sailed on the *Terror* in the 1840 Ross expedition. Reaching the bay, they found it to be packed with ice. It looked so thick that Scott was concerned that not only would the *Discovery* fail to break through, but that, if she did so, she could become trapped. Besides, any attempt to ram their way through would be akin to slamming into granite, risking damage to the hull. Not yet willing to take any unnecessary risks, he decided to keep heading east.

Sailing past the volcanos Erebus and Terror, named by Ross after his two ships, they continued eastwards to come alongside the Great Ice Barrier. With its coastal ice wall towering at over 200 feet, this tremendous floating ice platform is the size of France, with 90 per cent of its gigantic mass lying beneath the water. Unsurprisingly, such a barrier is impossible for any ship to penetrate.

Coasting east along this wall of ice, Scott planned to reach Ross's furthest point of exploration before then travelling on into the unknown. In doing so, the *Discovery* set her first record of the expedition, having sailed further south than any man had ever done before, passing 78 degrees 30' south. Soon after, another momentous discovery was made.

Having taken over a week passing alongside the Great Ice Barrier, they caught sight of a rocky island that couldn't be located on the map. In his excitement, Shackleton recorded, 'It is a unique sort of feeling to look on lands that have never been seen by human eye before.' Scott named this new discovery King Edward VII Land, after the monarch who had wished the party well before departing from the Solent. Despite this, the *Discovery* kept moving east, still on the lookout for a suitable base.

Time was now growing short. Autumn was approaching and the pack ice was becoming ever denser, meaning the *Discovery* was increasingly in danger of becoming trapped. Scott now realized that moving ever further into uncharted waters, with no guarantee of finding a base, would be foolhardy. He therefore decided that it was time to utilize the hot-air balloon.

Spotting a small inlet that ran 12 miles into the Great Ice Barrier itself, Shackleton, Scott and others took the balloon ashore. Yet with Scott concerned that the waiting ship might have to make a quick getaway, should encroaching ice be about to trap her, he ensured they took emergency stores with them so that they would not be stranded if the worst was to happen.

Quickly assembling the balloon, they hoped that the unparalleled views it would offer would reveal a place to set up camp. At the very least, they would be able to see further across the unexplored terrain than any man before. Having been trained to set up the balloon, and been responsible for finding the funding for it, Shackleton hoped he might become the first man to take flight in the Antarctic. But to his disappointment, and with no time for a joy ride, Scott took charge.

In the face of a strong wind, and with the balloon tethered to the ground only by a single wire, this was extremely dangerous. If the wind were to dislodge the balloon from the wire, then there would be no saving whoever was on board at the time. Wilson felt it was a risk not worth taking, grumbling that it was an 'exceedingly dangerous amusement in the hands of such inexperienced novices'. Scott felt he had no choice. Time was running short and he felt this was the most efficient way to look for a place to set up camp. He nervously clambered on board, and the balloon soon reached a height of over 600 feet. If he had been nervous before he was soon dazzled by the sight of the limitless white horizon. Yet as awe-inspiring as the view might have been, there was still no obvious place to set up base.

As Scott descended and considered his options, Shackleton took the opportunity to climb into the balloon. Carrying a camera with him, he took the first photographs of this vast undiscovered expanse of snow. At the very least, he had this to cheer himself. However, a leak was soon found in the balloon and Shackleton's time in the air came to a swift end, as did the use of the balloon. The inlet subsequently earned the imaginative name Balloon Inlet to commemorate the short-lived experiment.

Setting sail soon afterwards, Scott ordered an immediate return to McMurdo Sound, hoping this time to find a route through the ice. Approaching tentatively, where there had once been a thick belt there was now an open stretch of water. Not wasting a second, Scott ordered the ship onwards, soon anchoring, in Wilson's words, at the 'most perfect natural harbour imaginable'. Not only was it in a protected nook, it also looked like it had access to the snowfields of the Barrier and, hopefully, to the South Pole itself. In the perilous circumstances, it was all they could ask for. Shackleton was, however, a little unnerved, describing the bay as having a 'weird and uncanny look'. Nevertheless, this alien world would be his home for the rest of the winter.

Scott now had to decide whether they would live in a prefab hut on land or anchor the ship overwinter and live on board. He decided on the latter, especially as pack ice would soon solidify comfortably about the ship, keeping her in place.

While some were delighted to have the comfort of the ship for the winter, Scott's decision did not sit well with everybody. Some sailors had hoped that the ship would leave the shore party behind on McMurdo Sound and that they, and the ship, would head back to New Zealand, returning the following summer to pick up Scott and his crew. Now, they faced months trapped on the ice.

However, for Shackleton, there was nothing but excitement. He had feared being sent back with the ship to New Zealand but now had the chance to overwinter and experience something few men had before him. And with the South Pole just 900 miles away, there was still every chance that he would make history, and his name.

8

Everything had to be in order before the sun disappeared for the next four months, plunging this strange new land, and the *Discovery*, into darkness. While they would overwinter on the ship, the three prefabricated huts still needed to be erected on land and filled with supplies, in case of emergency, at what would become known as Hut Point. Scott also wanted to use the quickly diminishing light to explore the Barrier and to try to find a route south which they could take when they later set off on their true adventure.

Scott had, however, hurt his leg during ski practice and decided that one of his ship's three officers should lead the patrol in his stead. A coin was tossed and, to Shackleton's great delight, he won. Aged twenty-eight, he was to command the first sortie to locate a route to the South Pole. Life could hardly get better. This was just the type of adventure that had drawn him to the expedition in the first place, but not everyone was quick to offer their congratulations.

Some of the crew were bemused that such a vital role had been decided on the toss of a coin. The jocular Irishman had never before set foot on the ice for any prolonged period, let alone led an expedition, no matter how small it might be. There were others on the crew who seemed far more suitable for the role, particularly Armitage, Bernacchi and Koettlitz, who had all been on previous polar expeditions.

Despite the grumbles, Shackleton selected his two closest friends, Wilson and Hartley Ferrar, a twenty-three-year-old geologist, to join him. On 19 February 1902 the trio set off into the 'weird white world', without dogs or skis, all the while hauling an 11-foot sledge which contained provisions for three weeks, as well as a collapsible dinghy, which would be needed if the ice broke up. On the back of the sledge, fluttering in the increasingly bitter wind, each man flew their personal banner, Shackleton's emblazoned with his family motto, *Fortitudine vincimus* ('By endurance we conquer').

Panting and wheezing as they dragged their sledges over the ice, the temperature dropped, with freezing winds howling around them, making for an eerie and unsettling scene. Just as in the Sahara, there is virtually no rain in the Antarctic, so comparatively little snow ever falls. As such, blizzards are the equivalent of sandstorms in the desert, with the gales peppering the men's faces with shards of ice rather than sand.

In the midst of the biting cold, the likes of which none of them had ever experienced, they huddled deep into their Burberry cloth. Some have criticized Scott's use of this item of clothing, feeling it didn't sufficiently protect the men from the elements. Instead, many believe that Scott should have provided Inuit animal skins and fur clothing for such a journey. However, while ideal for travel with dog-drawn loads, skins and fur are impractical for man-hauling. For that, clothing needs to be fully breathable, allowing body sweat to escape rather than to remain and dampen underclothes, which can freeze during rest halts. In addition, sweating can lead to dehydration, and without insulated vacuum flasks to hold adequate water, exhaustion may soon result. In the circumstances, the Burberry cloth was probably the right choice.

Trudging through the ice, shielding their faces as they did so, they found hauling the sledge far harder than any of them had envisaged. Stopping in frustration, Ferrar exclaimed that they were travelling only at a rate of 6 inches per step. Undeterred, Shackleton roared them on, his beard hairs tugged by ice, his eyelashes freezing together. Indeed, he soon learned some harsh yet necessary lessons when it came to operating in such extreme temperatures. Whenever he looked through his binoculars, he realized he had to avoid breathing on the lenses; if he did so, his breath would freeze on them and obscure the view. He also soon learned that wiping the lenses with his bare finger could bring on frostbite.

After twelve hours, the men could take no more. Exhausted, their muscles burning, and sweat freezing to their skin, Shackleton suggested they set up camp to get some rest. But even this turned into an ordeal. Frostbitten fingers and numb hands made it difficult to erect the tent, especially with the fabric being blown by the increasingly strong winds. As Wilson recalled, they were, 'Hanging on like grim death to the tent pole to prevent the whole bag of tricks going to blazes.'

When the tent was finally up, they all scurried inside to find some warmth and inspect their various injuries. However, their hands were so cold they could barely bend their fingers, taking several attempts to light their Primus stove. At last, with it lit, and after a much-needed drink of hot chocolate, the men fought to remove their primitive boots, to which their socks had become fixed with frozen sweat, before shuffling into their reindeer-skin 'sleeping suits' (no sleeping bags had as yet been unpacked) to try to get some sleep. This was, however, to prove difficult. 'The cold of the ice floor crept through,' Wilson wrote, 'and the points of contact got pretty chilly. We put Ferrar in the middle and though that was the warmest place, he was coldest during the night. I think he was most done up.'

Finding it impossible to keep warm, and the tips of their noses freezing if they poked outside their sleeping suits, the men woke constantly to bang their feet together, trying to ward off the danger of frostbitten toes. I well remember that the real ordeal in such conditions was needing to go for a pee in the night and having to leave the sanctuary of your sleeping bag. I learned the hard way that it was better to keep the pee bottle in your sleeping bag rather than brave the cold then have to warm yourself up all over again. Although it must be said that ensuring all of the pee went into the bottle with your body shivering was quite a challenge. If you missed the mark, the urine-soaked clothing not only turned to ice but also stank. Mind you, after a couple of months without washing we stank anyway, so we lived with it.

Even when in your sleeping bag, and with all the proper equipment, it is so cold at night that you continue to shiver, resulting in an average loss of 2,000 calories a night. By the time you wake up, you are ravenous but often have only meagre rations to look forward to. Doing this day after day can be quite debilitating, and again this was something that Shackleton would also have to learn to counter when quantifying his rations in marches to come.

Having had a chance to thaw out and get some rest, the next day a determined Shackleton soon had the party out on the glacier and making for the islands. In the back of his mind, he no doubt wanted to prove to Scott that his trust in him to lead this expedition was warranted and so didn't want to waste any time. Hauling onwards, they got to within 2 miles of the nearest island, but could not go any further. Exhausted, they once more set up camp before commencing the big push the next day.

After eight hours' sleep, Shackleton roused the men. With the islands so close, he suggested they leave most of their equipment behind and make for the island with the highest peak. However, Shackleton and the men found that their lack of experience and specialized equipment made for a poor combination. In the ever-increasing cold they found it difficult to grip with their fingers, while their footwear was also unsuitable for such a climb. Often crashing to the floor, they were all somewhat battered and bruised when they finally reached the top. At least the view that awaited them was worth it.

From 2,700 feet, they set eyes on more of the Barrier than had any man before them. Amidst the never-ending ice was a chain of hulking mountains way off in the distance. And way beyond that was the South Pole itself. This was indeed a moment to treasure. He, Ernest Shackleton, as Scott's chosen scout, had found a way by which the South Pole could perhaps be reached. If nothing else, he felt he had already made his mark.

With their mission accomplished, Shackleton and the men could hardly wait to return to the *Discovery*. Invigorated by their achievement, they quickly descended the mountain, collected their equipment and set off for the ship, making far quicker work of the journey on their return.

On board, Scott and the crew listened in awe as Shackleton regaled them with tales of what they had seen. Revelling in being the centre of attention, he held most of the crew rapt with tales of their adventure. Scott could tell that they were 'bubbling over' with enthusiasm, while First Lieutenant Royds wasn't perhaps as enamoured with the scout, claiming that Shackleton was 'full of talk, as was expected'. However, while Shackleton didn't stop talking until everybody had turned in for the night, Skelton noted, 'I will do him credit to say that he talks sense and not blooming nonsense like one of our members, if not two or three!'

Excited by Shackleton's initial report, Scott sent another party on a voyage of discovery just two weeks later. This time, their destination was Cape Crozier, which was to serve as one of the prearranged 'post' boxes.

On this occasion, the twelve men, led by Royds, took eight dogs with them to help share the load. But with the dogs untrained and conditions worsening, Royds had no choice but to order twenty-four-year-old Lieutenant Barne to return to the ship, along with eight of the men. It was to end in tragedy, with George Vince, a naval rating, slipping and plunging to his death into the icy sea below.

In shock, Barne soon lost control of the party, with the group breaking off into different directions. If it had not been for naval seaman Frank Wild taking control of three of the men, then far worse would have followed. Wild led them back directly to the ship, but it would take Barne, and three others, hours to find their way, finally arriving in a very sorry and incoherent state.

Scott was distraught, saying that the loss of Vince was 'one of our blackest days'. The men's inexperience in such conditions was proving to be telling. On 4 March 1902, Scott noted:

> I am bound to confess that the sledges when packed presented an appearance of which we should afterwards have been wholly ashamed, and much the same might be said of the clothing worn by the sledgers. But at this time our ignorance was deplorable; we did not know how much or what proportions would be required as regards the food, how to use our cookers, how to put up our tents, or even how to put on our clothes. Not a single article of the outfit had been tested, and amid the general ignorance that prevailed, the lack of system was painfully apparent in everything.

On such an adventure into the relative unknown, this was all only to be expected, although Bernacchi wasn't quite as forgiving. Having been a member of Borchgrevink's *Southern Cross* Expedition, he had more experience than anybody there of the Antarctic and believed that vital lessons were not being learned. Of particular concern was that many of the crew could not even undertake simple tasks, such as putting up a tent, lighting a stove or even dressing adequately.

Thankfully, day after day on the ice saw these amateurs slowly but surely picking things up, such as how to adjust their clothing, how to sleep with their socks or mitts close to the body, as well as the uselessness of the three-man fur sleeping bags. When first used, these bags weighed 45 lbs, but twenty days later, owing to the accumulated weight of frozen perspiration, they weighed 76 lbs, adding further weight to be carried.

Any further adventures on the ice would, however, have to wait. By March, the *Discovery*'s anchorage had fully iced over, with tons of ice squeezing against her from all sides, causing the tortured ship's timbers to creak and shriek. Then, on 23 April, the sun disappeared below the eastern horizon, plummeting them into 123 days of darkness.

9

When cocooned in the darkness for months on end, many on Antarctic expeditions have lost their minds. This is no surprise, especially for those experiencing it for the first time. It is not only dark, cold and claustrophobic, but being marooned on a mysterious land thousands of miles away from civilization can be difficult to get your head around. These pressures would take their toll on even the most hardened of individuals. Indeed, on the *Belgica* expedition, the crew succumbed to a mix of scurvy and 'cabin fever'.

Even those on the *Discovery* who kept their sanity found the conditions difficult. As temperatures outside dropped to as low as minus 62° Fahrenheit, most decided to stay on board, but even then, ice still formed on the cabin walls. In these uncomfortable conditions, many of the men grew homesick, while being trapped in the same place and seeing the same people, for months on end, quickly began to grate. Any natural tendency to irritation, depression, pessimism or worry was magnified and, in the absence of loved ones, the natural, indeed the safest, outlets for pent-up feelings were diaries and letters home. That's human nature, and even more so on an expedition where sensibilities can become extreme.

I found on my first expedition south that forced togetherness breeds dissension and even hatred between individuals and groups. Because of this, at times, I found it very difficult to interact with my fellow teammates Oliver Shepard and Charlie Burton. Some days, without a word being spoken, I knew that I disliked one or both of the other men, and I could tell that the feeling was mutual. I was, however, very fortunate that at base I had Ginny to keep me company, who was always a tremendous comfort and support. With my wife by my side, I could at least let off some steam to her; failing that, I would vent my spleen in my diary. It is quite funny now, looking back, as some of my gripes were so inconsequential, but at the time they felt like major disagreements.

In those dark winter months, with the wind howling outside, Clarence Hare, Scott's steward and the youngest man on board, revealed there were frequently quarrels and fights among the crewmen. This was where Shackleton truly came into his own.

In charge of the ship's entertainment, he had ensured that the library was stocked with books while he helped put on plays and even arranged debates, going head to head with Bernacchi over the poetic merits of Browning and Tennyson, with Shackleton arguing for Browning. The crew eventually voted for Tennyson, albeit by a single vote.

Perhaps Shackleton's greatest contribution to keeping the crew occupied and entertained was the *South Polar Times*. In the sanctuary of an 'office' that was boxed off in the hold, Shackleton sought to publish this journal once a month, inviting the crew to submit any scientific articles, stories, poems, drawings or jokes. Wilson's illustrations featured heavily, as did Barne's caricatures of the crew, while Shackleton always found time to contribute a poem, written under the pseudonym Nemo, the hero of Jules Verne's *Twenty Thousand Leagues under the Sea*. Calling one of his poems 'To the Great Barrier', in it he alluded to uncovering the mysteries of Antarctica:

> This year shall your icy fastness resound with the
> voices of men,

Shall we learn that you come from the mountains?
  Shall we call you a frozen sea?
Shall we sail Northward and leave you, still a Secret for
  ever to be?

There were also plenty of other things to keep the crew entertained. In the privacy of a small cabin, the men could 'sport the oak' (contemporary public-school terminology for masturbation), and food was always something to look forward to. This was particularly true for the officers, as Scott did all that he could to make mealtimes an event, with fine china and engraved cutlery being set out in the wardroom.

Shackleton himself could also be relied upon as a great source of entertainment, always eager to share stories or a joke. A deckhand wrote, 'He was a very nice gentleman, always ready to have a yarn with us.' As had always been the case, he could mix with anybody and was liked by most.

One of the crew that Shackleton especially enjoyed spending time with was twenty-seven-year-old Frank Wild. After eleven years in the Merchant Navy, Wild had enlisted in the Royal Navy just a year previously but, craving adventure, had signed up for the expedition. In this respect, he was very similar to Shackleton and it was easy to see why they got on so well. But it was still with Wilson that Shackleton spent most of his time. After a meteorological observatory had been set up on nearby Crater Hill, both men volunteered to regularly take the readings, even if it meant climbing the hill in the darkness and bitter cold. Shackleton explained, 'We want to do as much exercise . . . as possible, and this will be a useful way of so doing.' This no doubt allowed them some welcome time away from the ship, from which the always patient Wilson had previously complained 'there is absolutely no escape', but it also had a far greater purpose: for

Shackleton to prove to Scott that he should be selected as part of his team to go south when the summer finally commenced.

Places to join Scott on his expedition south were highly coveted; it was for this reason that many of the men had signed up in the first place. There was no one more willing than Shackleton. He certainly was not there just to make up the numbers. From the time he told tall tales to his sisters about becoming a famed hero he had dreamed of an opportunity like this. Now, it was within touching distance – as long as Scott selected him.

When selecting team members for my own polar expeditions, I always tried to choose confident, highly capable individuals who, although not 'yes' men, would not threaten my status as leader. It's a difficult mix. While you need people who can think and act quickly under pressure, you don't want to have to deal with any attempted mutinies on the ice. Sometimes, it's impossible to tell how people will react until they're tested in some of the harshest conditions known to man.

Scott, therefore, had a difficult decision to make, particularly as he favoured taking only one other person with him. In this he had been inspired by Fridtjof Nansen, whose famed north polar wanderings had been undertaken with only one companion. With just one place up for grabs, Armitage was the strong favourite. He had actually been promised a place on the main southerly journey right from the outset, hence why he took the position as Scott's second-in-command. However, as time had passed, Scott had slowly but surely grown to depend on the solid character of Edward Wilson. Having already proven himself on the ice, Wilson was also intelligent and good company. Furthermore, he was a doctor, which could prove vital on such a hazardous journey.

In mid-June, and having had a chance to observe all likely candidates during the winter months, Scott informed Wilson that he would be travelling south. Wilson was stunned. He had only joined the expedition to study the wildlife and geology. Indeed, he had no real inclination to go south, as he revealed in a letter to his wife, complaining that the journey would be 'taking me away from my proper sphere of work to monotonous hard work on an icy barrier for three months'. There was also something of a question mark surrounding Wilson's health. While he was now apparently well, he had initially been turned down for the *Discovery* expedition, as he had suffered a bout of tuberculosis that left him with permanent lung damage, something that could prove debilitating on a long trek in freezing conditions. In the circumstances, Wilson felt there was any number of people who could have been chosen ahead of him. However, in spite of all the reasons not to go, he could not turn the opportunity down. 'My surprise can be guessed,' he wrote. 'It is the long journey and I cannot help being glad I was chosen to go.'

Wilson might have been bemused by the decision, but Shackleton was devastated. Although Shackleton was quick to pass on his genuine congratulations to his friend, he became downbeat and withdrawn. Suddenly, the one real chance to make a name for himself, to be etched forever into the history books, looked to be over. All that would remain was further months marooned on the ice, away from Emily, waiting for Scott and Wilson to return. It was a demoralizing prospect. But Shackleton wasn't out of the running just yet.

Wilson was concerned that, with just the two of them out on the vast Barrier, if one should be incapacitated, it could prove fatal to the other. In private, Wilson therefore suggested to Scott that perhaps he might consider taking

Shackleton as well. After some thought, Scott realized that Wilson's reasoning made some sense. Shackleton had already proven his worth on the glacier, and he was a powerful man, capable of man-hauling for hundreds of miles. He was also resourceful and good company. It was therefore settled. Come October, the three of them would set off, aiming to make history, hoping to, at the very least, break the furthest south record of travel towards the pole.

After a number of polar expeditions with one companion, and others with two, I look back and can see that things were always easier with just one other person. With only two of you, each time the leader makes a decision that the other person disagrees with, the latter will usually chew the matter over in their mind and will usually do as they are told. But with three, if two disagree with the leader's decision, they might discuss it among themselves and an awkward mini-revolt may result. Even if this fails, it can lead to a bad atmosphere. If I lived all my expeditions over again, I would never take a third party if it could be avoided (unless one member of the group was my wife). As the old saying goes, 'Two's company, three's a crowd.'

For the beleaguered Shackleton, this was a tremendous stroke of good fortune. He was of course 'overjoyed' when Scott asked him to become the third man for the great trek south, although, once again, not everyone shared his cheer. When the news broke there was a considerable amount of discontent from some of the crew. They could understand Scott taking Wilson, because he was a doctor, but they felt that the other place should surely have gone to a naval man. Armitage was particularly angry, writing bitter things about Scott for many years to come. Skelton also found it hard to bite his tongue, writing, 'Shackleton gassing and eye-serving the whole time . . . of course the Skipper's ideas are on the whole perfectly right . . . but . . . why he listens to Shackleton so much beats me – the man is just an ordinary gas-bag.' Yet no matter what anyone thought, Shackleton was Scott's man and his decision was final.

With final preparations underway, and August bringing increasing hours of sunlight, Shackleton was tasked with attempting to train the dogs on the ice. This was a tall order; Shackleton had no experience, and the rowdy and undisciplined dogs were especially difficult to tame. Neither whipping them nor friendly persuasion worked particularly well, and Shackleton was too impatient to persevere for long with either.

Some critics (including Scott in his own diary) suggested that he should have started his dog-team trials earlier than he did, even doing so under the winter moonlight. But such a course could easily have led to severe frostbite and dead dogs, as Amundsen's Norwegian team was to experience some years later. Starting dog trials at the end of August, near his base, as Scott did, was undoubtedly the most sensible timing. Although he was not totally convinced by their performance, Scott eventually concluded that the dogs made things slightly less difficult than no dogs.

In the meantime, Armitage was at least given a consolation prize for missing out on the journey south. On 11 September 1902 he set out with five men to find a route to the Magnetic Pole, where Earth's magnetic field lines come vertically out of the surface. Finding this location would be crucial to the updating of magnetic maps of the southern hemisphere and thus to marine navigation. In contrast, the geographic South Pole, which Shackleton hoped to reach, is the place where all the lines of longitude converge in the southern hemisphere. For their journey, Armitage and his team all took skis, a decision about which a still-sceptical

Scott commented, 'I am inclined to reserve my opinion of the innovation.'

However, as Armitage led his men to an inlet named New Harbour, two of them, Ferrar and Heald, became very ill. Closer investigation revealed that they were both suffering from an affliction that would threaten the success of the expedition: scurvy.

Scurvy caused bruising, weakened muscles, could lead to tooth loss, severe joint pain and, in the worst cases, death. Today, we are well aware that scurvy is the result of a lack of vitamin C. However, this wasn't the case at the turn of the last century. Certain patterns had been picked up over time, particularly that men near shore and eating fresh fruit, vegetables and meat seemed to be immune from the disease. It seemed most common when men were on long voyages, away from these things. Once they returned to the coastline and ate a heavy diet of fresh seal and penguin meat, they miraculously recovered, but no one yet realized that the crucial element was vitamin C.

First Officer Charles Royds was particularly puzzled by the men's illness, commenting, 'When one thinks of the fresh meat which has been constantly provided throughout the winter . . .' While they had eaten a rich diet of seal and penguin meat, they did not know that when the meat was cooked or dried the amount of vitamin C in it was diminished. As such, it is likely that Heald and Ferrar were already severely deficient in vitamin C before they set out with Armitage.

Back at the base, Wilson and Armitage did what they could to care for Ferrar and Heald, while also ensuring that the other men avoided a similar fate. Ordering more seal-hunting trips, Armitage also told the cook, Henry Brett, to be a bit more imaginative in the meat's preparation, as many of the men disliked the taste. Now, with seal served every day, rather than two or three times a week, Heald and Ferrar quickly recovered, while the crew on board remained in good health.

At this time, Scott, Shackleton and Thomas Feather, the boatswain, were away on another mission before the big push south. Carrying with them a sledge-load of provisions, helped along by sixteen dogs, they aimed to lay supply depots out on the Barrier which they would use when the march south started for real in October.

Although scurvy had been an issue on Armitage's journey, Scott's depot-laying operation had also raised some health issues. Of particular concern was Shackleton, with Wilson recording that he 'has blistered all his fingertips again with frostbite and is obviously a bit done up. He is very much thinner and has lost pounds in weight.'

Wilson was aware that the two short expeditions that Shackleton had endured on the ice so far would pale in comparison to what lay ahead. If they were to reach the pole, as they hoped, this would entail a round trip of 1,480 nautical miles, man-hauling 16 miles a day for over one hundred days in the most testing conditions known to man. Despite having recommended Shackleton to Scott, Wilson now wrote of his concerns to his wife, Oriana: 'I feel more equal to it than I feel for Shackleton. For some reason I don't think he is fitted for the job. The Captain is strong and hard as a bulldog, but Shackleton hasn't the legs the job wants. He is so keen to go, however, that he will carry it through.'

Wilson decided against telling Scott of his concerns, even though Shackleton was now suffering from regular coughing bouts that kept him awake at night. It was clear that he wasn't well, but he didn't want to miss out on the opportunity of a lifetime. In an indication that Shackleton had his own concerns about his health, the night before departure he wrote a letter to Emily, saying that if he were to die on the trip, 'not to grieve for me for it has been a man's work and I have helped my little mite towards the increase of knowledge . . . Child, we may meet again in another world . . .'

There was also something crucial that had been forgotten by all and should have now rung alarm bells: Shackleton was the only individual on the ship who had not passed a medical exam before setting sail.

10

It had been nine years since Markham had first dreamed of a British expedition to conquer the unknown south. And now, after two years of preparation by Scott, the big day had at last arrived. On the morning of 2 November 1902, the chosen trio of Scott, Wilson and Shackleton posed for a photograph alongside their nineteen dogs and five sledges. The sledges were loaded with provisions and equipment to last a total of seventy days, so, no matter what, they had to turn back after thirty-five.

Before departing, Scott left orders for his second-in-command, Albert Armitage: 'Should I not return before a date at which there is any possibility of the ship being again frozen in, you should take the ship back to New Zealand after provisioning the hut and leaving, if you think fit, a search party which could be recovered the following season.'

In hooded jackets, their faces protruding from a wire-rimmed funnel, designed by Scott to protect them from frostbite, the three men took their first steps towards history, to the cheers and well wishes of their compatriots on board the *Discovery*. While the dogs pulled the sledges behind them in a train, a support group consisting of Barne and twelve other men were already out on the Barrier, having left three days earlier to lay provision depots for the trio's return journey.

As the *Discovery*, Mount Erebus and Mount Terror slowly disappeared from view, early attempts at skiing when pulling alongside the dogs proved unhelpful. However, after the third day the snow surface changed and on the flat terrain the skis worked well, leaving the men able to travel an average of 11 miles a day for four days straight.

With the dogs also performing well, the three polar knights soon caught up with Barne and his depot-laying team, who had travelled without skis and had found the going tough. Despite this, the team had still set a furthest south record of 78 degrees 55'S, shattering Borchgrevink's record. This certainly gave Scott, Wilson and Shackleton hope that they could do even better, especially given their superior equipment and preparation.

Sadly, the dogs' performance now began to wane. Their feed, a mix of Norwegian dried codfish and biscuit, had been soiled during the *Discovery*'s journey south, leaving them badly undernourished and struggling to complete the trek's daily target. Neither were they eating any fat, which would have provided them with the required fuel for such a cold climate. The animals certainly knew what they were craving, as one night one of the dogs managed to escape and eat a week's worth of fatty seal meat from the men's rations. Gradually slowing, and their reserves of energy diminished, Wilson began to call the dogs 'sooners' – they would 'sooner' do anything than pull.

Their rapidly deteriorating performance was also partly due to the absence of any visible object ahead of the lead dog. In 180 degrees of featureless waste, there was nothing to aim for, which meant they were confused and slowed down. This is a problem that has also badly affected me, but for very different reasons. When I was suffering from double vision after a long expedition, a top eye surgeon diagnosed that the long weeks of travel with no horizon and no real focus point had made my eyes focus at a distance of one metre ahead of my boots. It took weeks for this 'convergence of the visual axes' to relax.

To counter the problem with the dogs, Scott ordered each of the men to take turns hauling alongside those at the front. This was a mistake, as it merely encouraged the dogs to take it easy and let the humans do all the work. Moreover, on the uneven terrain, with three men having to haul a heavy load of over 490 lbs between them, their skis were rendered useless. Worse still, each pair weighed 20 lbs, so, lashed to the sledge, they were a substantial addition to the dead weight that now needed to be hauled.

During my expeditions, it has also been my experience that any benefits skis might have are virtually lost the moment you start to haul a heavy load, which is why it baffles me that so many critics continue to castigate Scott, and others, for not putting more faith in skiing.

In 1993, when Mike Stroud and I became the first to cross the Antarctic landmass unassisted, we each man-hauled a sledge with a start-weight of 495 lbs, often without the help of skis. For context, this was the equivalent of lashing together three average-sized adults, each weighing 160 lbs, dumping them in a fibreglass bathtub with no legs and then dragging it through sand dunes for 1,700 miles, for ninety-six days. It was back-breaking work, but we achieved our goal, proving to the critics that skis and dogs are not vital for non-mechanical travel in the Antarctic.

That is not to say, however, that it is easy work, particularly for those with no prior experience of man-hauling in such conditions. Although Wilson complained of sore knees and tight hamstrings, it was Shackleton who really concerned him. 'Shackle started a most persistent and annoying cough in the tent,' Wilson wrote, adding a little while later, 'All of us very fit . . . except Shackle, whose cough seems very troublesome.' This certainly didn't bode well, particularly at such an early stage.

The men now began to slow even more. Hungry, tired and short of breath, the weather conditions didn't help matters either. However, it was the heat rather than the cold that was causing the problems. With the wind dropping and temperatures rising to minus 6.5° Celsius, hauling was even harder work, particularly when the men were clothed for sub-zero conditions and were therefore sweating profusely. On one such occasion in the Antarctic, it was so hot I had to man-haul in nothing but my long johns and vest. Yet due to the ozone hole, which at least wasn't an issue for Scott, Shackleton and Wilson, I soon found that my skin was burning. When I did stop, the sweat particles that had dripped down my underwear and on to my nether regions froze, which was a memorably painful experience.

In such conditions, and with the men struggling, it became clear to Scott that he would have to change tack if they were going to reach their goal. He therefore took the painful decision to divide and relay the load. Now, they would pull half the load for 5 miles (or less in bad visibility), depot it, return for the other half, then continue on. This meant that they were travelling 3 miles to gain a single mile of progress. Scott also switched from daytime to night-time travel, in order to avoid the hottest time of day, since, despite twenty-four hours of daylight, the sun was a lot lower at 'night'.

I have done the same many times during my expeditions, particularly on a solo journey to the Antarctic, when I had no choice but to split and relay a 470-lb load in order to ascend Frost Spur from sea level to the polar plateau, having lost my crampons down a crevasse. It took five back-breaking trips, but there was no way I could have carried the load up that steep and icy slope all in one go.

While Scott was right to relay the load, Shackleton still described it as 'Trying work, this going backward and forward' and, later, 'The travelling is awful.' Even the brief periods of rest offered little respite, as their sweat froze on their skin, making them even colder.

Each night at camp the men tried to thaw out, relishing their pemmican 'hoosh' (a dried meat mixed with fat then melted with a hard biscuit) to warm their insides. Amidst the howling gales outside, the darkness was made all the more frightening by the sound of 'quakes' caused by the collapse of a suspended mass of ice which sparked a thunderous roar across the landscape. To take their mind off such things the men came to rely on each other, with Wilson reading passages from the Bible, and Shackleton telling stories or reading poetry.

The three were an unusual mix, with Scott from the Royal Navy, Shackleton a merchant seaman, and Wilson a doctor, but they all got on well. Indeed, even their diaries, which were written in private and were an opportunity to let off steam, did not betray any dislike for each other. However, it seems that one incident did cause momentary resentment.

One day, when making the hoosh, Shackleton accidently spilt it on to the tent floor, burning a hole in the groundsheet. Scott snapped at Shackleton for not taking more care, but it appears that the incident was soon forgotten, no doubt helped by Wilson, who was always a soothing presence. Neither man mentioned it again, and neither did they record any words against each other in their private diaries afterwards. In my opinion, the fact that no one said a detrimental word about the other only goes to show how inconsequential this event was, despite some biographers trying to make much more of it.

Far from admonishing Shackleton, Scott could only applaud his efforts. Struggling with shortness of breath, Shackleton nevertheless continued onwards, his whole body bent forward, as far as it would allow, to give him the strength to haul the sledge with the harness on his shoulders, encouraging the others as he did so. Scott said of him, 'In spite of his breathless work, now and again he would raise and half-turn his head in an effort to cheer on the team.' By hook or by crook, Shackleton was determined not to let anyone down, least of all himself. Pulling the sledge with all of his might, on 25 November, they passed the 80th parallel, the nearest anyone had got to the pole from any part of Antarctica's circumference.

Still, they were falling behind schedule. Scott realized that in order to gain more days for southerly travel, he would have to cut their daily rations to less than 29 oz each. For breakfast, there was now nothing more than a mug of chopped bacon and biscuit, while lunch was more biscuits, boiled seal chunks and a mug of hot chocolate. At night, they fared a little better, eating a mixture of pemmican, bacon, biscuit and cheese, along with a protein additive called plasmon. Yet this was not nearly enough to keep them trekking miles upon miles in treacherous conditions while hauling a heavily laden sledge. Daily, they were now only consuming around 4,000 calories but probably using more than 7,000 calories. At this rate their weight loss would have averaged around 10 oz a day and, since most of their body fat would have disappeared in their first month of travel, by December their starved bodies would have been metabolizing the very muscle they depended upon. Moreover, when you are hungry your body starts to live off its fat stores, and in breaking the fat down the liver produces chemicals called ketones, which circulate in the blood and can make you feel generally unwell and depressed – a volatile mix alongside pain, exhaustion and starvation.

In my view, it was this calorie deficit, more than anything to do with the use of skis and dogs, which was to prove the greatest challenge for the men. Through no fault of his own, Scott had no idea how many calories they would need to consume, as he did not know how testing the landscape would be.

For our 1993 crossing of the Antarctic continent, Mike Stroud and I aimed to consume 5,200 calories per day, but even that was not enough. After measuring our calorific expenditure, Mike, Britain's number-one professor of research into stress nutrition, found I was burning 10,670 calories a day! Analysis of our blood samples showed that our enzyme systems, everything that controlled our absorption of fat, were changing, and we had gut-hormone levels twice as high as previously recorded. We were adapting to our high-fat rations in a way hitherto unrecognized. Furthermore, with zero remaining body fat, we were losing muscle and weight from our hearts as well as our body mass.

Something similar no doubt happened to Scott, Shackleton and Wilson. Soon the men thought of nothing but food, as their hunger grew and grew, like some living, gnawing creature in the depths of their stomachs. In his diary, Shackleton titled a page 'Desire' and wrote below a list of foods, which included sirloin steak, crisp fried bread, jam pastries and porridge. Every recorded dream also involved eating some sort of food. In one of them, he dreamed that 'fine three-cornered tarts are flying past me upstairs'. Scott, meanwhile, wrote that the men had now turned from 'the hungry to the ravenous'. And things soon got worse. Fuel stocks were running low so Scott had no choice but to instigate the economy measure of chewing frozen rather than boiled seal meat chunks at their midday halts.

As they marched onwards, battling tiring legs and increasingly empty stomachs, the dark smudges of distant land at times seemed close, and then mists would swirl about them, blocking out the views. As Wilson noted, 'The coast we are making for is still about 50 miles away . . . It looks very beautiful though, all snow-covered peaks, bold cliffs and headlands.' Then, soon afterwards: 'Snow grains falling all day. Nothing but "white silence" all round us.'

With nothing at all to steer by, they attempted to follow a compass bearing, a slow and frustrating task even after weeks of experience. When the compass failed them, with no visible horizons, white-out conditions, zero perspective and no aiming mark, Scott often had nothing more to rely on than a shred of wool tied to a light bamboo pole so that he could be steered by the direction of the wind.

When my time came to navigate across the frozen continent in the late seventies, I had to use exactly the same techniques as Scott and Shackleton. Satnavs and GPS were still unusable at such latitudes, as there were no polar orbiting satellites at the time. I spent weeks being taught by the army to use theodolites, sextants, chronometers, nautical almanacs and sight reduction tables, plus many nights learning to recognize individual stars in each quadrant of the sky. Like Scott, at times I also had to improvise. When the sun shone, I sometimes navigated by means of the shadows of a series of penknife-scratched lines on the plastic windshield of my skidoo. The ability to adapt and improvise is definitely the key to success in such testing conditions, and Scott was proving to be a master of it.

By mid-December, the endless relaying, the intensifying hunger, tiredness, dropping temperatures and the ailing dog pack saw daily progress down to just 2 southward miles. Shackleton's diary included entries such as 'Fingers very sore', 'Legs tired', 'Feet tired', 'Dogs not pulling . . . Must struggle on' and 'Tired with hauling.' All that was keeping Shackleton going was sheer willpower, but even that was beginning to falter.

To preserve energy, talk was kept to a minimum, while the skis were discarded to lighten the load, as were further provisions, left behind at Depot B. I recall doing the same when trying to cross the continent with Mike Stroud, deciding I could dispense with my duvet jacket, as I had hardly worn it. Disposing of such items is of course a calculated risk. At the moment you decide to get rid of anything that is dead weight, you pray that you won't come to miss it, as I did my duvet jacket, when the temperature plummeted only days afterwards.

Scott's group had travelled more than 100 miles in just over a month, but with the relay work they had probably travelled three times as far, with raw heels now grinding against their boots, each step sending flashes of sharp pain across the men's taut features. They were all now nearing the end of their endurance. Even the usually positive Shackleton knew they were close to staring defeat in the face, commenting, 'We cannot go on any more like this.'

It rapidly became apparent that further sacrifices would need to be made if they were to continue. In order to stop the soul-destroying relay of sledge loads and to use all of their energy on marching forward more weight would have to be lost. Now, they faced the hardest choice of all, as Wilson wrote, 'We have decided now to feed up eight or nine of the best dogs on the others, and when these dogs have eaten each other up we must pull our own victuals and gear.' It was a decision that left all of the men cold, none more so than the animal lover Scott, but it would also mean that they would no longer have to haul dog food, as the dogs would eat each other: truly, it was 'dog eat dog'.

Some have suggested that Scott would never have premeditated such un-British dog slaughter, even though he knew that the Norwegians accepted it as a sound means of polar travel. However, although Scott was unable to force himself to drive dogs to death in the manner described by the likes of Fridtjof Nansen, he had always accepted the potential need for dog slaughter. The previous winter, along with Wilson, he had even pre-planned the precise mathematics of progress by the dog-eat-dog system, as can be seen in Wilson's diary. He must have hoped that they would never have to use it, but he was ready for it should it prove necessary.

Trudging painfully south, two men man-hauling alongside the remaining dogs, and one cracking the whip and shouting from the rear, they slowly grew to loathe the experience and silently vowed never to repeat it. As each dog died, or was killed by Wilson, who pierced the heart of the animal with his medical scalpel, the men recorded the dogs' deaths in their diaries like the passing of friends. When Wilson was unable to continue with this gruesome task after suffering with snow-blindness, Scott refused to take over. Ashamed of his squeamishness, he wrote, 'It is a moral cowardice of which I am heartily ashamed and I know perfectly well that my companions hate the whole thing as much as I do . . . [Wilson does] all my share of the dirty work.' It was therefore left to Shackleton to do the deed, but the bloody sacrifice was at least worth it.

Having lightened their load, the three men were now able to get back to travelling at least a mile an hour. It was still too slow in the great scheme of things but far better than their previous efforts and, now that relaying had been abandoned, at least all the distance covered was forwards. Nevertheless, the dream of reaching the pole seemed to be slipping away, particularly as each man's health took a downward turn.

Snow-blindness now afflicted them all, Wilson particularly

badly. In one of his bloodshot eyes, he felt as if the eyeball was being repeatedly stabbed. In trying to describe his ordeal, he wrote, 'My eye was so intensely painful that I could see nothing and could hardly stand the pain. I cocainized it repeatedly. I never had such pain in the eye before . . .' In the seventies, while on a training expedition in the Arctic, I was to suffer with something very similar. With a windchill factor of minus 120° Fahrenheit, the natural liquid in my eyes actually congealed. It was sheer agony.

There were times during the journey when two out of the three men were blind and the third had the use of only one eye. To protect his ailing eyes, Wilson dragged himself, and his load, along blindfolded, while Scott described to him the view, even though it was more often than not just relentless whiteness stretching as far as the horizon.

As if this wasn't enough, with the men not getting anywhere near enough vitamin C, scurvy now reared its head. Shackleton was the first to show early symptoms when his gums became inflamed, and days later Scott and Wilson suffered in the same way, along with fatigue, painful joints and aching muscles. With no fresh vitamin C available, and being unaware in any case that this would be the cure, the three men had no hope of avoiding the next stage of the disease: more painful muscles and bones, loose teeth, haemorrhaging and gangrenous ulcers. If they didn't get any vitamin C soon after that, death would follow. But where could they get such a thing when out in the wilderness?

Thankfully, Christmas Day at last offered the men a short reprieve. Scott loosened restrictions on rations and they feasted on seal's liver and bacon for breakfast, followed by a hot lunch. As if to show just how important a good meal was, the men then proceeded to march further than they had in weeks, walking up to 11 miles, something to be celebrated that night with a mug of hoosh and a special surprise, courtesy of Shackleton.

Having hidden a plum pudding in a spare sock, for this very occasion, he proceeded to present it to the wide-eyed Wilson and Scott. At that very moment he was not only the most popular man on the continent, but maybe on the face of the Earth. Spirits were the highest they had been in weeks as the three men eagerly tucked in, leaving not a crumb.

On the back of a good day's march, and with full bellies, the men huddled in their tent and spoke about the days ahead. They were within 10 miles of 82 degrees south and some 490 miles from the pole. They had already set a new furthest south record and were aware that their food and fuel stocks would allow for only four more days' outward travel. It was therefore decided that they should try to pass 82 degrees south by 28 December and then head back to the *Discovery*. As Shackleton wrote, 'It will not be sage to go farther.'

Despite the men being debilitated by snow-blindness and scurvy, Shackleton was determined to lead the men past the 82-degree parallel. Out in front, he hauled the sledges and encouraged Scott, Wilson and the surviving dogs onwards, pumping his tired legs and arms through the snow. Thanks to his extraordinary effort, on 28 December, they reached their goal, passing the so-called 'magic circle', but as Scott recorded, in the process, 'We have almost shot our bolt.'

The 82-degree parallel put them in the mouth of a wide west-running inlet between two sets of mountains. While they had intended to turn back, they realized that, with just a few more miles of southerly toil, they might be able to see sufficiently far into this great inlet to learn whether it was merely a bay or a channel cutting off the coast. Such information might prove vital for future expeditions in reaching the pole. Deciding to make one last effort south, Scott wrote:

> We argued, however, that one never knows what may turn up, and we determined, in spite of the unpromising outlook, to push on to our utmost limit. As events proved, we argued most wisely, for had we turned at this point we should have missed one of the most important features of the whole coastline.

Shackleton was, however, unable to continue. He desperately needed to rest after putting everything into reaching the 82nd parallel. A decision was made to leave him behind in a tent while Scott and Wilson went ahead with his blessing. On the last day of 1902, they set off, skiing into the inlet, and with the fog and the clouds dispersing, confirmed that, since there was no visible end to it, the 20-mile-wide inlet must be a strait between two islands. Scott subsequently christened it Shackleton Inlet as a consolation to his friend. He also named a nearby mountain Mount Markham, in honour of the founder of the expedition. In setting out on this final journey, Wilson and Scott set a furthest south record of 82 degrees 17'. Shackleton, back at camp, had only reached 82 degrees 15'. At the very least, despite all their troubles, they had beaten the furthest south record by more than 200 miles.

Upon Scott and Wilson returning to Shackleton in the tent, they all agreed that they had gone as far south as their bodies and provisions would allow. In the circumstances, and being one of the first teams ever to attempt such a crossing, it was a valiant effort. Yet the pole was still hundreds of miles away, a seemingly impossible conquest. Scott might have felt a 'deep sense of disappointment' at not getting any closer, but as they turned back Shackleton wrote that it was 'a wonderful place and deserves the trouble to get there'. Indeed, they had discovered and mapped nearly 300 miles of virgin coast.

However, if they didn't return safely to the *Discovery*, no one would know of the sights they had seen or of the record they had set. Once more they would be put to the test of human endurance, none more so than Shackleton, who soon became so ill he was unable to stand.

11

The group now faced a race against time. Rations were low and they had to make their way back to Depot B, where they had unloaded provisions – over 100 miles away. But finding this tiny spot was not an easy task. Marked by a single black flag, in the vast white Antarctic landscape, the depot could have easily been buried by drifting snow, or a howling gale might even have dislodged the flag. Moreover, their compasses did not indicate an accurate magnetic variation, making their task even more difficult. Yet if they were delayed, their meagre rations would run out and they would surely die. As Wilson put it when they headed back north on 31 December 1902, 'We *must* reach the Depot . . . before January 17th.'

For four days, the exhausted little group strained every muscle to average 8 miles a day. If they could keep this pace up, then they should just about reach the depot in time. But by 5 January the men were forced to do all of the work, with most of the remaining dogs walking behind, an all but useless canine caravan, surviving on cannibalism as their numbers dwindled. In normal circumstances, they might have considered eating the dog meat to satisfy their great hunger, and even, perhaps, to help stave off scurvy, but this was not considered advisable. The dogs were sick and the men were therefore concerned that their meat was tainted.

They were now slowing alarmingly and there was still no sight of the depot. Quick flashes of hope were repeatedly dashed, and while no one yet said it, there was real concern that they would not make it back. But then, at midnight on 13 January 1903, as Scott scanned the horizon through the telescope of his theodolite, he was thrilled by the sight of a tiny black spot in 360 degrees of whiteness. 'I sprang up,' he wrote, 'and shouted, "Boys, there's the depot." We are not a demonstrative party, but I think we excused ourselves for the wild cheer that greeted this announcement.'

That night they camped at the depot, where they had stored fourteen days' worth of provisions, aiming to make the most of the rest, and the food. However, it would have to keep them until they reached Depot A, which required them to march 7 miles a day, and now with the added load of the skis which they had picked back up at this second depot.

If the dogs on the latter stages of the march south had been labelled 'sooners', then the remaining animals, in Wilson's words, were now 'only a hindrance'. Something that certainly didn't help matters was the party having opted to carry some of the ailing animals on their sledges, increasing the weight still further. Such foolishness clearly had to end, with a distraught Scott ordering the remaining dogs to be killed. 'I think we could all have wept . . . [at] the finale to a tale of tragedy,' Scott wrote. 'I scarcely like to write of it.' Neither he nor Shackleton would ever forget what they had been forced to do to their dogs. They would remain determined never to repeat the experience.

Faced with such unrelenting bleakness, one innovation at least looked to propel them forward. Using the floor cloth of their tent as a sail, they harnessed the strong winds, hoping they would push them all the way to Depot A. Sadly, their makeshift kite was blown in the wrong direction, and soon they were lost. As Scott wrote: 'We cannot now be far from our depot, but then we do not exactly know where we are.'

I recall having a similar issue. When attempting to cross

the Antarctic continent with Mike Stroud, we used parachutes, rather than tent cloths, to try to propel us forward. This was lethal. The wind was so strong we were dragged through belts of sastrugi, ridges of ice cut by the wind, at dangerously fast speeds, with our sledges frequently toppling over. Sometimes our sails caught on ice obstacles, snapping us to a halt, or became hopelessly tangled, leaving us to try to untangle them in the freezing wind. In the end, Mike and I decided they were far too dangerous to persevere with.

However, in 1996, when I attempted to become the first man to travel solo, unassisted, across Antarctica, I did so with the aid of a kite. My Norwegian rival, Børge Ousland, had sung the praises of this method, and with it I was able to harvest the winds and travel 117 miles a day while not exerting anywhere near as much energy. In fact, I conserved so much energy I struggled to eat all of my 1,600 daily calories and actually dumped eight full days of rations, thinking I wouldn't need them. Alas, while I was making record time, I was struck down by an attack of kidney stones and had to abandon my attempt. In any event, Børge Ousland was to claim the record, making the crossing in a remarkable fifty-five days, having sailed for three quarters of the distance.

Now lost in white-out conditions, and with their food supplies dwindling, Shackleton, Scott and Wilson desperately searched for the depot's black flag. However, with their visibility obscured by mist, and the featureless Barrier stretching for hundreds of miles, it became all but impossible to locate it. As Scott said, it was a 'very small spot on a very big ocean of snow'. It seemed they were in need of a miracle to find this proverbial needle in a haystack. Thankfully, they received one.

With a brief break in the mist, Scott saw what he thought looked like a black flag on the horizon. Although the mist closed in seconds later, Scott was sure of what he had seen. Without a moment to lose, they quickly marched in the direction of where the flag had been spotted. This was their last hope. It was truly all or nothing. If they reached the spot and it turned out to have been nothing more than a mirage, they were in serious trouble. To their relief, it had not been a trick of vision. It was indeed Depot A, and, with it, apparent salvation.

As they triumphantly wolfed down a hot bowl of hoosh soon after reaching the depot, it was, however, clear that Shackleton had taken himself to his limit. Wheezing heavily and coughing up blood, he was in a very sorry state. Although the hoosh allowed him to continue marching the next day, Wilson truly feared for the well-being of his friend, writing that he 'has been anything but up to the mark'.

The *Discovery* was still over 150 miles away, yet Shackleton's deteriorating condition left him unable to pull his sledge. Without complaint, Wilson and Scott took on the extra burden, watching in concern as the ailing Shackleton trundled next to them in clear distress. 'Captain Scott and Dr. Wilson could not have done more for me than they did,' he wrote. 'They were bearing the brunt of the work and throughout the difficulties and anxieties of such a time showed ever cheery faces.'

Unsurprisingly, the extra load was slowing the men down, at a time when every minute was vital to their survival. All but essentials were now dumped on the ice and the last two dogs slaughtered. This lighter load helped Scott and Wilson somewhat, but now Shackleton was finding it too difficult even to walk.

Violently coughing up blood, Shackleton collapsed on 18 January with pains in his chest. Scott and Wilson hurriedly set up camp so that he could rest, while Shackleton apologized profusely, conscious that his ill health was costing them all dearly and feeling humiliated at being reduced to a passenger. Scott recorded that Shackleton, 'feels his inactivity very keenly'.

A dramatic last roll of the dice soon gave Shackleton half a chance of making it back to the *Discovery*. With just one set of skis remaining, he opted to give them a try. Thanks to strong winds, and with minimal effort, he found that he could keep up with Scott and Wilson. After travelling over 10 miles one day, and the party making it to another depot, the smoking stack of Erebus signalled that they were now just 100 miles from Hut Point. But one look at the emaciated Shackleton told Scott that even travelling on skis and a full stomach might not be enough to save him. He wrote, 'There is no doubt Shackleton is extremely ill.' Wilson's medical opinion, meanwhile, was that he was 'quite unfit' to travel.

Such was Wilson's concern that, as Shackleton drifted in and out of consciousness, he told Scott that he might not survive the night. Years later, Shackleton revealed that he had heard Wilson say this to Scott and it had given him a renewed determination to prove them wrong. Against all the odds, the bull-headed Irishman rose the next day and, with a little help, staggered on to his skis and drove onwards.

Despite this display of astonishing willpower, it was not to last. Soon tiring, Wilson persuaded his friend to sit on the sledge so that they could tow him. Shackleton knew he had little choice, but the hurt to his pride was devastating.

Dragging Shackleton on, Scott and Wilson were also now finding the going desperately difficult. Wilson's legs burnt, leading to a heavy limp, while Scott's ankles became badly swollen, all the effects of weeks of treacherous marching and scurvy taking their toll. 'We are as near spent as three persons can well be,' Scott confided in his diary.

On 3 February, staggering onwards for mile after mile,

getting ever closer to the ship, the outline of two men suddenly appeared. They thought it must be a mirage, or their addled minds playing tricks on them, but as conditions cleared they saw that it was Bernacchi and Skelton, who had come to greet them. They were saved. However, the sorry sight of the men certainly took Bernacchi and Skelton by surprise. 'They appeared to be very worn and tired,' Bernacchi recalled 'and Shackleton seemed very ill indeed.'

After such an ordeal, it was no surprise that the bearded, malnourished and frostbitten men looked as they did. Over the course of ninety-three days they had marched 960 miles, man-hauling sledges for much of the way, across the world's most treacherous continent. In the process they had also answered to some extent one of the territory's great mysteries. Rather than a floating mass of ice, Antarctica was really a frozen land mass covered almost entirely in ice sheets.

Upon arriving back at the *Discovery*, which was still trapped in ice but with supplies recently replenished by the relief ship *Morning*, the 'three polar knights' were greeted with three cheers from men clinging on to the rigging. Helped on to the ship by Koettlitz and Royds, the weakening Shackleton immediately turned in to rest, after enjoying his first bath in ninety-four days. While he later roused himself to join the crew for a three-course meal of soup, mutton and plum pudding, he was so exhausted he could not even finish the first course.

Shackleton was proud of all he had achieved. Now he was back on the *Discovery* he felt he could return to Emily as a hero, a record breaker no less, his name being made. But things would turn out not to be quite so simple.

12

Although the march south had been completed and a new record set, the *Discovery* could not yet depart. Still encased in the ice, the crew would have to wait a few more months for it to thaw. That's not to say there still wasn't much to do. There were still a number of scientific studies to complete, not to mention further sledging journeys to explore the continent.

In the meantime, Shackleton took the opportunity to regain his strength, eager to play as full a part as possible in the remaining months. He was still somewhat embarrassed at being relegated to a passenger on the return journey and very much wanted to prove his worth.

However, not far from the *Discovery* was the *Morning*, Markham's relief ship. While its primary purpose had been to provide fresh supplies, as well as deliver mail to the crew, it would also be able to take any unfit men, or those who simply wanted to leave the expedition, back to New Zealand. Shackleton certainly had no plans to be on the relief ship home. Indeed, despite his poor condition, when Scott asked if there were any volunteers to return, he kept his hand firmly down. Despite this Scott remained concerned about Shackleton's health and asked Koettlitz, the senior doctor, to examine him.

Koettlitz found no permanent damage and, perhaps responding to the patient's clear desire to stay, told Scott that he felt Shackleton need not go. Scott wasn't convinced. He had issued a directive that all officers on the ship should 'enjoy such health that [he] can at any moment be called upon to undergo hardships and exposure'. He therefore pressed Koettlitz for a definite opinion on whether Shackleton would be fit to lead a sledging journey, to which Koettlitz wrote, 'Mr Shackleton's breakdown during the southern sledge journey was undoubtedly, in Dr Wilson's opinion, due in great part to scurvy taint. I certainly agree with him; he has now practically recovered from it, but referring to your memo as to the duties of an executive officer, I cannot say that he would be fit to undergo hardships and exposure in this climate.' In a letter Koettlitz wrote to Scott Keltie at the Royal Geographical Society, he also disclosed that Shackleton 'had been attacked by some sort of asthma'. Although Shackleton was well on his way to recovery from scurvy, this underlying condition was clearly still a worry.

Wilson also remained concerned for his friend. He had witnessed Shackleton's rapid deterioration during their return journey, as well as his continued breathing difficulties and coughing fits. In his mind, it was inadvisable for Shackleton to risk a second winter. Weighing all of this up, Scott told a dismayed Shackleton he would return home on the *Morning*.

Shackleton felt humiliated. Not only had he had to be pulled on the sledge on the final leg of their journey, now he would have to leave the ship, discarded as an invalid. Rather than feel proud of all he had achieved and return home a hero, he now felt he was anything but. Bernacchi wrote that Shackleton was 'deeply disappointed and would give anything to remain'. So devastated was he by Scott's decision, Shackleton begged Reginald Ford, who was in charge of records and accounts, to swap duties with him, so he might be of some use in a less taxing role. Ford turned down Shackleton's request and, in any event, it was unlikely that Scott would have sanctioned it.

When word swept the decks that the popular Shackleton was to be sent home, there was much disappointment. Aware of the furore, Scott insisted that it was purely on health grounds. 'It is with great reluctance that I order his return,' he said, 'and trust that it will be made evident that I do so solely on account of his health and that his future prospects may not suffer.'

Whatever he might have hoped, Shackleton's expedition was now at an end. In his eyes, everything had been in vain, and it was all down to Scott. From here on in, it has been claimed by many, a feud between the two began, although some have claimed it had begun far earlier.

In the years that have followed, there has been much written about this perceived feud, most of which initially came from the mouth and the pen of Albert Armitage. While he wrote a book about his experiences on the *Discovery* two years after the event, in it he failed to mention any rift between Scott and Shackleton. However, as the years passed, a terrible marriage and the strains of service at sea saw Armitage grow increasingly bitter, particularly towards Scott for having overlooked him for the journey south.

In 1922, when Hugh Robert Mill was writing his biography of Shackleton, Armitage produced a bombshell. He revealed:

> Shortly after their return from the southern journey, Shackleton told me that Scott was sending him home, and asked me if I could do anything about it. He was in great distress, and could not understand it. I consulted Koettlitz and he informed me that Scott was in a worse condition than Shackleton. I then went to Scott and asked him why he was sending Shackleton back. I told him there was no necessity from a health point of view, so after beating about the bush he said, 'If he does not

go back sick he will go back in disgrace.' I told Shackleton and promised to look after his interests.

Armitage then added another ancient 'memory', also told for the first time:

During the winter, Wilson told me the following story, which Shackleton confirmed later. On the southern journey, Wilson and Shackleton were packing their sledges after breakfast one morning. Suddenly they heard Scott shout to them: 'Come here you BFs [bloody fools].' They went to him, and Wilson quietly said, 'Were you speaking to me?' 'No, Bill,' said Scott. 'Then it must have been me,' said Shackleton. He received no answer. He then said, 'Right, you are the worst BF of the lot, and every time that you dare to speak to me like that you will get it back.' Before Shackleton left he told me that he meant to return to prove to Scott that he – Shackleton – was a better man than Scott.

Would Wilson, Scott's great friend and renowned for his loyalty, have told such a story to Armitage, with whom he was never close? The polar historian David Wilson, Edward Wilson's great-nephew, has no doubt that Armitage, embittered by the passage of two difficult decades, invented the conversation.

Indeed, notes written by Skelton, the chief engineer on the *Discovery*, also cast doubt on Armitage's reliability. 'Armitage is a peculiar chap,' he wrote, 'especially with regard to his arguments . . . His methods are not always genuine.' After Armitage's death, Skelton was even more forthright in his views:

I think Armitage's character was often rather soured – he did not think he got enough credit for his work – he was not

> popular in our Expedition nor in his own Service, the P&O. Scott was tactful with him and none of us had rows. He had that silly inferiority complex of the Merchant Navy for the Royal Navy. In his private life he had much misfortune. His brother in the RN, who I knew well, committed suicide rather than face a court martial for a nasty crime. His wife was 'a hell of a woman' and poor old Armitage, whom she made penniless, was far too kind to her. The last ten years of his life after leaving the P&O found him very impecunious and embittered by lack of appreciation by others . . . Personally, I never had a row with him but never corresponded or got near to him and that goes for all of us on the *Discovery*. His books are not always accurate, and rather poor stuff . . . he is inclined to claim too much.

Why should Armitage have invented such a story? Some believe it was due to Scott inadvertently asking him if he too wanted to leave on the *Morning*. What Armitage didn't know was that Scott had been given confidential instructions from the Admiralty, via Markham, that he should try to persuade Armitage to return. The reason remains unknown but, in the murky whispers, it appears to have involved Mrs Armitage's young child and a brewing scandal with some third party. Scott's attempt to put this tactfully to his second-in-command misfired badly, and Armitage decided that he was to be purged merely because he, like Shackleton, was Merchant Navy. He later wrote, 'Fortunately my appointment was independent of [Scott]. I absolutely refused.'

Another myth encouraged, if not invented, by the embittered Armitage was his 'memory' that Wilson once told him he 'had it out with Scott' during their southern journey and that 'it' involved Scott's treatment of Shackleton. In 1933 the author George Seaver quoted, apparently from papers which

Wilson's widow let him see before she destroyed them, the same single sentence from Edward Wilson: 'Had it out with Scott.' Some Scott biographers have used this ambiguous Wilson quote, of which no written source exists, to show that Wilson confronted Scott about his treatment of Shackleton. There is no evidence that this was so.

There is no mention in Wilson's meticulous diaries, or anywhere else for that matter, of any incident that this could conceivably have referred to. Yet this slender tale cited by Seaver has been seized on by generations of polar historians as evidence that Wilson, acting in his habitual role of peacemaker between all men, attacked Scott over some specific incident that occurred between Scott and Shackleton or, perhaps, to warn Scott against aspects of his behaviour. Maybe Scott's oft-confessed bad moods, impatience and quick temper were the topic of this tête-à-tête? No one will ever know; they can only speculate.

However, what needs no speculation, since Wilson, Scott and Shackleton's diaries, and subsequent letters, are all available to the public, is that nowhere, in any of them, is there any mention of any rift between Scott and Shackleton at this time. I have mentioned previously that when writing in your private diary during such testing times you are inclined to let off some steam. For Scott, a man known to disappear into dark moods and hold a grudge, it is remarkable that, if there was any issue with Shackleton, he did not write about it.

Indeed, Scott's letter back home to his mother truly reveals that he bore no ill feeling towards Shackleton. 'All the crocks I am sending away,' he wrote, 'and am much relieved to get rid of. Except Shackleton, who is a very good fellow and only fails from the constitutional point of view.'

From my own experience, the acid test of how men have fared with each other on such a long and nightmarish journey comes at the end, when it is at last possible to escape from each other's company. The night after the three men returned to their cabins on the *Discovery*, Gerald Doorly, one of the officers from the relief ship *Morning*, was passing by the neighbouring cabins of Scott and Shackleton, when he heard Scott call out, 'I say, Shackles, how would you fancy some sardines on toast?' This hardly supports the rumours of any estrangement, let alone hostility, between the two men.

Moreover, at the time when Scott sent Shackleton home, the seventy or so men on the *Discovery*, and on the relief ship, were mingling and gossiping, as expedition members are wont to do, trying to find out from one another the inside story. Those team members due to stay in Antarctica for another year subsequently wrote in their diaries, or to their loved ones back home. Many of the men's diaries, and many of those letters (including fifty-six letters written by the garrulous Charles Royds) have been scrutinized again and again in later years by Scott biographers searching for the slightest mention of the rumoured rift between Scott and Shackleton. None mentioned any ill-feeling whatsoever. The only inference, out of all the men on board, that there might have been more behind Scott's decision than medical grounds came from the biologist, Thomas Hodgson, who wrote, 'I hear it is true, as I suspected, that personal feeling is the real reason for Shackleton's departure.' But how would Hodgson have known this? It certainly would not have come from Wilson or Scott so we can only attribute it to Chinese whispers.

The passage of time has long since buried the truth, but in sending Shackleton home Scott wrote to the Royal Geographical Society, 'Mr E. H. Shackleton, who returns much to my regret, should be of the greatest use in explaining the details of our position and our requirements for the future.'

If Scott truly had any issues with Shackleton, then I do not believe he would have trusted him as his spokesperson for the expedition until he returned in person. In my mind, it is clear that Scott was so concerned for Shackleton's health, and knowing all he had already achieved, that he thought it best he return home. This of course was not much consolation to Shackleton.

On 1 March, an unsteady Shackleton left the *Discovery* under suitably grey skies. As he sombrely walked across the ice towards the *Morning*, his compatriots, among whom he had made himself so popular, scaled the rigging and waved their goodbyes, sorry to see him go. 'I cannot write much about it,' Shackleton wrote, 'but it touched me more than I can say when the men came on deck and gave me 3 parting cheers.'

A day later, the *Morning* left the *Discovery* behind and set off for New Zealand. Shackleton, a proud man, who had dreamed of making his reputation, and fortune, cried. He felt he had let himself and Emily down. She would not be expecting his return for another year, and now he would be home much earlier and, in his eyes at least, a failure. Whatever the truth behind Scott's reasoning to send Shackleton home, he had ensured one thing: that until the day he died the two men would be rivals.

13

After docking in Christchurch, Shackleton made the long journey back to Britain via San Francisco. He was naturally not in the best of moods, as can be discerned from his less than charitable diary comments about his fellow travellers.

'There are a fair number of passengers on board but they seem to me to be a pretty dull and uninteresting crowd,' he grumbled to Emily. 'There is one individual or rather two that I have made up my mind not to sit next to: for coming up in the train they did nothing but growl and talk about themselves. "More! More! About yourself!!!"'

Of course, after adventures such as his own, other topics of conversation might have been somewhat dull in comparison. He was hardly in the mood to talk in any event. Rather than try to entertain others with swashbuckling stories of his escapades, for now, at least, it was all still too close to the bone. The gnawing feeling of failure and resentment continued to eat away inside him.

After almost two years away, Shackleton finally returned home in June 1903. By now he was at least starting to resemble his old self, having had a chance to collect his thoughts, rest and enjoy some hearty meals. Despite his own feelings of having failed, he was relieved that Emily still saw him as a hero, and certainly meant to marry him. However, one of the other reasons for going on the expedition was to make his fortune, in order to keep Emily in the style she was used to. He had even made this promise in writing to her father.

Now he had been sent home early and, having been paid a pittance while he was away, he was concerned that he was facing a financial black hole. Thankfully, as far as Emily was concerned, her late father had provided her with the sum of £700 per year (£40,000 today), so things were not entirely desperate. Indeed, this was three times as much as Shackleton had earned on the *Discovery*. Yet Shackleton found this humiliating. He not only wanted to provide for his future wife but to also ensure that he was the primary breadwinner.

The choices facing Shackleton were stark. He could return to sea, possibly with a Castle liner, but that seemed like taking a step back rather than the gigantic leap forward in prospects he had envisaged. After all, he was a record breaker who had endured conditions that had almost cost him his life. He was entitled to believe that perhaps there should be a reward for that. Rather than look back, he now sought to look for new opportunities.

Shackleton hoped his expedition experience might allow for an accelerated progress into the ranks of the Royal Navy. In relying on Markham and his extensive contacts, Shackleton felt sure that his application would be rushed through, particularly due to Markham's effusive praise for his role on the *Discovery*. 'Captain Scott speaks of Mr Shackleton as a "marvel of intelligent energy",' Markham wrote. 'He was never tired, always cheerful and is exceedingly popular with everyone. He retained Captain Scott's confidence and warm approval for his diligence and usefulness throughout the voyage and he accompanied his Commander on the memorable sledge journey to the South.'

Despite such effusive praise, Shackleton's application was rejected. While he might have blamed himself, he was not to know that the man Markham had approached on his behalf,

Sir Evan MacGregor, the most influential civil servant at the Admiralty, was no fan of Markham and was disinclined to do him any favours. As such, Shackleton's application was dismissed with barely a glance.

Another opportunity to return to sea did initially pique Shackleton's interests. With the *Discovery* still trapped in the Antarctic ice, the Admiralty sought to send a ship to assist her. As Shackleton knew the continent, and its conditions, far better than most, he was asked to serve as chief officer aboard the rescue ship *Terra Nova*.

It would seem Shackleton might have grasped this opportunity. In a dramatic role reversal, he would now be Scott's rescuer, having been sent home as an invalid. Indeed, an Admiralty doctor had passed him 'fit for Antarctic service'. Yet he turned it down. While there is nothing in writing that can help us deduce the reasons why, perhaps having to spend yet more months away at sea when he was planning a wedding caused him to think again. He did, however, travel to Dundee, to advise on the outfitting of the *Terra Nova* for Antarctic conditions, with her setting sail in late August 1903.

Soon after, Shackleton's Antarctic expertise was called upon again, when Otto Nordenskjöld's ship, *Antarctic*, was crushed by the ice of the Weddell Sea and sank. With the captain and crew stranded on a drifting ice floe, at the mercy of the conditions, it was a desperate situation. Eager to assist, Shackleton advised the international rescue party as best he could, with Nordenskjöld and his party all successfully rescued as a result.

Despite all of this, money-making opportunities were few and far between. Shackleton had hoped that employers would be beating his door down after his exploits, but instead there was silence. Desperately trying to come to terms with his perilous situation, he came to realize that he was in a unique position and that his prospective fortune was in fact staring him right in the face.

Shackleton had been appointed to act as the expedition spokesperson until Scott returned. It struck him that, with Scott likely to be trapped in the ice for the foreseeable future, he could set the narrative as he pleased. Thus, he could now talk himself up as the record-breaking hero and have all the acclaim, and opportunities, to himself. This would of course need to be done before Scott returned and told his own version of events. Without a second to waste, Shackleton made himself available for lectures and wrote articles for newspapers and magazines about his experience, earning much acclaim as he did so.

We already know that Shackleton was a good writer. After all, he had already written a book, as well as editing the expedition's monthly newspaper, the *South Polar Times*. With a natural storyteller's flourish, his articles on the expedition were therefore well received, although Bernacchi would later comment that he had 'managed to advertise himself extensively. From various accounts he does not seem to be quite playing the game.' For Shackleton, this was of course entirely the point.

While Shackleton told his story well in the written press, what really set him apart was his gift for public speaking. Always an enthralling storyteller, he now had a wonderful story to tell. Setting himself as the brave hero conquering an unknown land, audiences soon came flocking, and with this came some of the opportunities he had hoped for.

Heralded far and wide, his burgeoning popularity saw him approached by the Tabard Cigarette and Tobacco Company, who ran a shop in New Burlington Street, London. Eager to capitalize on Shackleton's growing name, he was invited to take a part of the business, with talk of potentially floating on the stock exchange. That dream never came close to fruition, but Shackleton was happy enough to have a minor interest in the business for years to come, for little or no work.

Still, Shackleton felt he had something to prove to himself, and perhaps even to Scott. He realized that to conquer these feelings of self-doubt he would need to go on another expedition. Keeping an eye out for the next possible adventure, he offered his services to Captain Joseph-Elzéar Bernier, a Canadian mariner, who was hoping to reach the North Pole in 1904. Sadly, the expedition never materialized.

With further doors being closed, and Shackleton's thirtieth birthday fast approaching, he turned to another skill he might be able to turn to full-time employment. Following the relative success of his book, and the *South Polar Times*, as well as his articles on the *Discovery*, he believed that he might make a good fist at being a journalist. It seems others thought the same, as in the autumn of 1903 he was appointed sub-editor at the *Royal Magazine*, a monthly journal aimed at providing light entertainment for the middle classes.

Shackleton could no doubt write, but it was soon clear that the role of a sub-editor was not for him. Rather than writing his own articles, he was in charge of editing and proofreading those of others, something that didn't suit a man who always preferred to be the protagonist. Bored and restless, he seemed to spend most of his time in the office regaling the fascinated staff with tales of his own adventures, which saw editor Percy Everett say, 'No man told a story better. He made us see things he spoke of and held us all spell-bound.' However, while Everett was enamoured with Shackleton, he also professed some misgivings. 'He was the most friendly "hail-fellow-well-met" man I have ever come across,' he said, but 'his knowledge of the technical side of bringing out a magazine was nil'.

Shackleton's journalistic ambitions came to an end after just three months. Being deskbound clearly didn't suit a man who craved adventure and enjoyed being his own boss. Perhaps a key factor in his decision was that the post of Secretary to the Royal Scottish Geographical Society (RSGS) had become available, a far more beguiling opportunity in his eyes. To his delight, he also found that he had an ally inside this institution who was prepared to grease the wheels to ensure he got the role.

John George Bartholomew, one of the founders of the RSGS twenty years previously, felt that the maverick Irishman with experience of a record-breaking expedition, no less, could be just the sort of dynamo that could bring the sterile RSGS into the new century. Both Markham and Hugh Robert Mill, Shackleton's friend from the *Discovery*, with whom he had continued to keep in regular contact, also gave him strong references. Indeed, such was Mill's loyalty and enthusiasm for Shackleton that he told the RSGS he would resign his membership if they did not give him the job.

In January 1904 Shackleton was overjoyed to be appointed RSGS Secretary, albeit on the meagre wage of £200 a year (about £11,000 today). The role involved moving to Edinburgh, with fiancée Emily by his side, whereby they rented a home at No. 14, South Learmonth Gardens. While he was thrilled at the opportunity, these were hardly the great riches he had promised his bride-to-be. Hoping to at least redeem himself in the future, he wrote, 'I only wish I could give you a lot beloved, but things will come on and soon we can do better', and 'my love is so strong that it will redeem the poverty of the rest of me'. Despite this, the RSGS job was at least viewed as prestigious, and Shackleton saw it as a necessary stepping stone to greater things. The position certainly offered dizzying possibilities, particularly as the people he would now be rubbing shoulders with, such as William Beardmore, one of the wealthiest businessmen in Scotland, in the words of Mill, 'held the keys to many locks'.

When, on 11 January 1904, Shackleton arrived for work in his rebellious light tweed suit, the sombre RSGS office on Edinburgh's Queen Street didn't know what had hit it. Previously, the office had been almost funereal, with many of the staff quite happy to be stuck in the past and ruled by outdated tradition. Indeed, Shackleton cruelly referred to one of the committee members as the 'Ancient Druid', while he said of Finlay, the interim secretary, 'tact and humour were out for a holiday when Finlay's life called for his share'. Determined to enact change, he hit the offices like a tornado, causing uproar when he chased local firms to *advertise* in the Society magazine, while his addition of a typewriter and a telephone was felt by many to be beyond the pale. Shackleton brushed such criticism aside. This was exactly what he had been brought in to do and he delighted in ruffling a few feathers. In writing to Mill, he said, 'You would have laughed had you seen their faces when the jangle of the telephone disturbed them.'

After giving a spellbinding lecture in Edinburgh's magnificent Synod Hall, Shackleton followed it up by instituting a highly popular series of annual lectures in major venues in Aberdeen, Dundee and Perth. Average attendance at these events rocketed from 250 to 1,630, and paid membership increased too, making a great deal of money for the RSGS in the process. If there were any misgivings about the rogue Irishman before then, these were now dismissed.

On Saturday, 9 April 1904, Shackleton finally achieved one of his great ambitions: marrying Emily Dorman. In Christ Church, Westminster, the happy couple, Shackleton now thirty and Emily nearly thirty-six, said their vows in front of friends and family. To Shackleton, it was the culmination of seven years of hopes and dreams. His broad face was as happy as if he had conquered the South Pole itself. Emily, meanwhile, had come to accept that the man she had married would always be a 'boy'. Despite his best intentions, she knew that a life of domesticity was not for him. She would never be able to chain him down, and did not intend to. As she later said, 'I never wittingly hampered his ardent spirit, or tried to chain it to the domestic life which meant so much to me.' Shackleton was extremely fortunate to find such a woman in Emily.

Honeymooning in the Highlands later that summer, the newly betrothed coupled enjoyed playing golf at Dornoch, with Emily gaining the upper hand. Soon after, and to their great joy, Emily discovered she was pregnant. It seemed that Shackleton had all he had ever wanted. As he said himself, 'I am so happy dearest thinking about all the times which are to be in the future.'

However, as with all explorers, once one goal had been met, then attention soon turns to the next. His job at the RSGS certainly helped to encourage dreams of further conquests, but finding a suitable expedition had proven to be tricky. None as yet had truly sparked much interest in him, with most not promising a chance for the glory he so craved. But what if he could lead his own expedition?

In July 1904, Shackleton met William Speirs Bruce, the veteran of five voyages to the Antarctic. Bruce had applied to join the *Discovery* expedition but had been turned down by Markham, no doubt because he was not a Navy man. Undeterred, Bruce had launched his own expedition to the ice, raising all the money required from private sponsors, many of whom were Scottish. On his expedition he had discovered many miles of new coastline, logged a remarkable 1,100 species and established Antarctica's first permanent weather station, on Laurie Island. While Shackleton was eager to trade stories about the continent, it was the fact that Bruce had led his own expedition, after being rejected by Scott, that really grabbed his interest. Why couldn't he do the same? He realized that he knew many of Bruce's donors and, just as importantly, they liked him. Better yet, they knew of his success with the *Discovery*, thanks to his series of lectures and articles.

With this thought whirring around in his head, news broke of the *Discovery*'s long-awaited return to Britain. It had been eighteen months since Shackleton had last seen Scott, and while there was not yet any sign of any true bitterness towards him, there was no doubt that he desperately wanted to prove him wrong, and leading his own expedition would do just that.

On 16 September 1904, Shackleton went to London to welcome Scott and the crew home. In front of cheering crowds, Scott was greeted like a national hero, promoted to captain and clearly enjoyed all the acclaim that Shackleton craved for himself. Watching Scott be showered with praise, Shackleton understood that his time in the spotlight was over. Scott was the golden boy now, and this made Shackleton even more determined to go it alone.

Over lunch, the two men tried to ignore the awkward past and instead focused on their future ambitions. Scott quite openly told Shackleton that one day he intended to return to the Antarctic and reach the pole. Shackleton wished him luck, but told him he had no plans to return, stating, 'I am married and settled down. I had thought of going on another expedition sometime but have given up the idea now as there seems to be no money about. It would only break up my life if I could stand it which Wilson says I could not.'

Despite this, Shackleton was already well on his way in making plans to return to the Antarctic, and hoping to beat Scott to the punch. As his old *Discovery* friend Hugh Robert Mill would later write, 'he had resolved at the time to prove that he was in no way unfit for polar service by going out again at the head of an expedition'. And this time, Shackleton would be the master of his own fate.

1. Ernest Shackleton (*standing at the back*) with his younger brother Frank (*seated*) and their eight sisters.

2. Ernest Henry Shackleton, aged eleven, at Dulwich College.

3. Shackleton joined the Merchant Navy and his first ship, the *Hoghton Tower*, in April 1890, aged sixteen.

4. Fir Lodge Preparatory School in Sydenham *c.* 1885. Shackleton is sitting at the back of the group in the window. He attended Dulwich College until 1889.

5. Scott's maiden reconnaissance balloon flight on 4 February 1902 after *Discovery*'s first landing on the ice barrier in the Bay of Whales.

6. Shackleton in London following his return from the furthest south record with Scott and Wilson. During the summer of 1909 his face appeared in almost every newspaper and his status as a national hero grew.

7. Shackleton modelling polar clothing for Burberry. Scott had selected Burberry cloth rather than Inuit animal skins and fur clothing for his *Discovery* crew.

8. The company of *Discovery* after their return to London in November 1904. Shackleton (*bottom row, fourth from the left*) joined them for the photograph, although he had not forgiven Scott for sending him home early.

9. Emily Dorman married Ernest Shackleton on 9 April 1904 at Christchurch, Westminster. She was thirty-five years old, six years older than her husband.

10. Cubicle at Cape Royds shorebase.

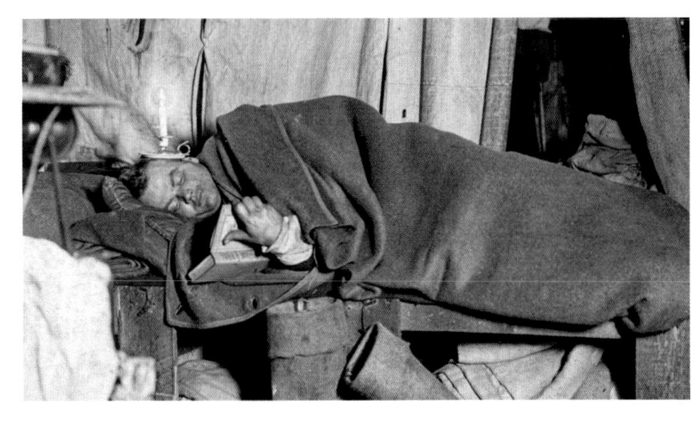

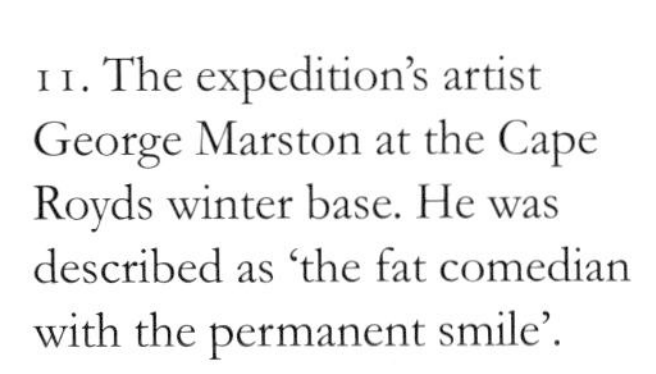

11. The expedition's artist George Marston at the Cape Royds winter base. He was described as 'the fat comedian with the permanent smile'.

12. A chess game helps pass time (a.) at Cape Royds and (b.) crossing the Antarctic continent in 1990 (Mike Stroud uses his blood samples as pawns). (c.) The Arrol-Johnson motor car, the first such car ever used in Antarctica, driven by Bernard Day for the first few miles of the southern journey while (d.) the four ponies each towed a sledge.

13. While Shackleton's party had covered 54 miles in the first nine days of their southern journey, bettering the 37 miles they had covered in the same period with *Discovery*, a fierce blizzard soon set in, and the party had no choice but to shelter in their tent and hope the storm passed. With every mile precious, Shackleton wrote, 'It is a sore trial to one's hopes'.

14. (*from left*) Wild, Shackleton, Adams and Marshall on board *Nimrod* after returning from their furthest south journey.

15. Clipping from the *Illustrated London News* of 12 June 1909. Shackleton surrounded by thousands of well-wishers welcoming him home, heralded as 'The Hero of the Moment'.

16. Ernest Shackleton, September 1908, still the hero who had been closer to the South Pole than any other. (Roald Amundsen and his Norwegian party would be the first to reach the South Pole on 14 December 1911).

17. Shackleton with Emily receiving a bouquet in Copenhagen during his 1909 European lecture tour.

18. Newspaper clipping from Shackleton's lecture tour.

19. Shackleton, always popular with women, holding court at a garden party, July 1914.

20. *Endurance* trapped in the ice, January 1915. Imperial Trans-Antarctic Expedition 1914–1916 (Weddell Sea party). Over the next ten months the ship moved with the floe, until October when ice began to crush the ship.

21. (a.) The *Endurance* crew playing football on the ice in 1915 and (b.) the author playing cricket at the South Pole in 1980.

22. A celebratory meal on Midwinter's Day, 22 June 1915, to herald the return of the sun. A tradition amongst Antarctic explorers.

23. (a.) Hussey holding Samson. This photograph is often captioned, 'The smallest man with the biggest dog'. (b.) Captain Scott took his terrier on board *Discovery*. The author's late wife Ginny had her Jack Russell terrier Bothy at both poles.

PART THREE

'A live donkey'

14

Being a part of an expedition, and leading one, are two very different things. You are not merely going along for the ride, focusing on your specific role, you are responsible for every single aspect, from recruiting the crew to buying a ship. And all of this takes time, particularly as you have to raise a tremendous sum of money. It took Ginny and me more than seven years to raise over £30 million in donations and sponsorship from 1,400 companies for our record-breaking Trans Globe Expedition. So, while Shackleton harboured hopes of leading his own expedition, he was about to learn some very hard lessons.

If his dreams were to become reality, he desperately needed to attract finance. Without it, the prospect of returning south was dead in the water. Yet after distributing a four-page initial prospectus of his plans to wealthy contacts, he didn't attract a penny of investment. This was a blow, but he was undeterred.

Considering who were the wealthiest, and most generous, potential benefactors, he struck upon William Beardmore, a shipping tycoon from Glasgow who was now turning his hand to the future: car manufacture. They had previously met at an RSGS function and Shackleton had also stayed at Beardmore's 200-year-old castle outside Glasgow. While he had initially ingratiated himself to Beardmore in the hope of securing a job in his sprawling empire, he now came to realize that this was just the man who could bankroll his dreams. Moreover, Shackleton also had a friendly, platonic relationship with

Beardmore's young wife, Eliza (who preferred to be known as Elspeth). With all this going for him, Shackleton was confident he could persuade Beardmore to back him.

In the meantime, another opportunity came like a bolt from the blue.

Shackleton's popularity, down-to-earth appeal and gift for public speaking saw the Liberal Unionist Party mark him down as someone who could have a future in politics. Shackleton agreed to meet with Sir John Boraston, the party agent and was asked to stand as MP for Dundee at the next general election. Shackleton quickly recognized that such a rôle would offer the prestige he had long dreamt of. There was just one problem; Shackleton did not have any firm political beliefs, or passion for the role. Perhaps Mill put it best when he described Shackleton's latest goal as a 'fine adventure and a tremendous lark', with Shackleton agreeing to stand without truly understanding what he was up against.

As far as Dundee was concerned, the Liberal Unionist Party was a lame duck. Having initially formed as a breakaway group from the Liberal Party, the Unionists were now the minority partner in the aristocratic and locally hated Conservative government. Moreover, the government's flagship policy was to oppose Home Rule for Ireland. At this time, Dundee had a large Irish population which had swarmed to the city for work. Such a policy was unlikely to win them over. In addition, Dundee was a working-class city of poorly paid millworkers often living in horrendous conditions who were fast aligning themselves to the emerging Labour Party, which promised to fight for their rights. The tide was therefore already turning heavily against the Conservatives, and in turn the Unionists, there, and across large parts of the country. In short, Shackleton stood next to no chance of being elected.

Still, Shackleton believed he had nothing to lose. He thought he could have a punt at being elected while also continuing in his role at the RSGS and all the while trying to mount an expedition. It seemed he wanted to have his cake and eat it, but it soon became clear that this was not an option. Due to a conflict of interests, the RSGS told Shackleton that if he wanted to run for Parliament, he would have to resign his position as their Secretary.

After weighing up his options, Shackleton decided that a life in politics was for him, telling Mill, 'With life before me, and strength and hope, all these things which time will whittle down, I may achieve something before the period at which life grows stale and strength wanes, and hope flies, or if it does not fly, assumes the dignified attitude of resignation.' In January 1905, he duly resigned from his position at the RSGS, where he had made such an impact. This was a monumental mistake. Not only did he stand little chance of winning, but no general election had yet been called. Indeed, the Coalition government was doing all it could to avoid calling one, knowing that if it did so, it would probably lose.

Without an election to fight, and on his way out of the RSGS, Shackleton became a father on 2 February of that year, when Emily gave birth to their son, Raymond. Proud though he was, claiming that his son had 'great fists for fighting', he now had another mouth to feed and was without a job, and it was all of his own doing. In a panic, he realized he had to find another source of income, fast, and put any thoughts of leading an expedition firmly on the back burner.

Again, journalism was an option. Shackleton was persuaded by Niels Grøn, a Danish entrepreneur, to invest £500 (£29,000 today) in his press agency, Potentia, which promised to tell the 'truth'. 'I feel it is going to be the great thing of the future,' he excitedly told Emily. But soon Grøn had squandered the money and not a single issue of Potentia was ever published. Another dream shattered.

While Shackleton had hoped to lead an expedition, he was now so desperate that he considered joining Michael Barne's expedition, which was looking to return to the Antarctic to explore Graham Land, an 800-mile northward extension of Antarctica towards the southern tip of South America, still relatively unexplored. But after Shackleton had discussed the idea with Markham, he found there was little interest in funding it. Once more, he was back to square one.

Perhaps the most upsetting blow of Shackleton's *annus horribilis* came in October 1905. Scott had finally published his book, *The Voyage of the Discovery*, and for Shackleton it did not make happy reading. While he had managed to paint a flattering portrait of himself on the lecture circuit, he was dismayed to read Scott's appraisal of him.

Scott wrote that Shackleton had been an 'invalid' who had been carried on a sledge, neglecting to mention that he had been suffering from scurvy, or indeed that he had bravely continued on skis when Wilson had thought he was close to death. Instead, it was made to look as if the trip had simply been far too much for the weakened Shackleton, with Scott declaring, 'I was the least affected of the party.' In the interest of public decency, Shackleton chose his words carefully. While he told Scott that the book was 'beautifully got up and splendidly written', he smarted from the criticism. It was as if a hot poker had been placed on a scar. Come what may, despite public appearances, he was determined to put Scott right. But before he could even think about doing such a thing there was, finally, an election to fight.

In December 1905, Balfour resigned as Prime Minister and an election was set for 16 January 1906. Recognizing this as a chance to save himself from further embarrassment,

Shackleton threw himself wholeheartedly into the election campaign, attending fifty-five political meetings in just three weeks. His captivating style certainly won over many voters, with the *Dundee Courier* praising his 'breezy personality and attractive manner'. Of particular note was his response to the question 'Would you advocate votes for women?' With a mischievous look in his eye, he lifted his fingers to his lips, and whispered, 'Hush! My wife is present.'

Despite the well-received wisecracks, Shackleton cut little swathe with the Irish population or the working classes. As an Irishman himself, Shackleton attempted to persuade them of the party's opposition to Home Rule by stating, 'I am an Irishman and I consider myself a true patriot when I say that Ireland should not have Home Rule.' Such a comment merely earned Shackleton the wrath of the hecklers in the crowd. Perhaps they could sense that he was being insincere, as his father was a strong supporter of Home Rule, as were his brother and sisters. In any event, he needed far more guile and intelligence to engage on such an inflammatory subject.

To no one's surprise, except perhaps his own, both Shackleton and the Liberal Unionists were annihilated in the election. Such was the Conservatives' fall from grace that Balfour became the first and only Prime Minister to lose his local seat. Shackleton came a distant fourth in Dundee, with just 13 per cent of the vote. The only person below him was the Conservative candidate. It was all so foreseeable, yet Shackleton had leapt into the fire without properly assessing the opportunity, so caught up was he in the quest for prestige and fortune. Bruised and somewhat humiliated, he at least retained his sense of humour, commenting, 'I got all the applause and the other fellows got all the votes.' Alas, applause did not pay the bills.

Having scrambled from one folly to the next in the three years since he had arrived home, Shackleton never gave up his 360-degree search. As the bitter and bloody Russo-Japanese War looked to be coming to an end, with over 200,000 having perished in just eighteen months, the humiliated Russians desperately needed help returning thousands of sailors from fighting in Southern Manchuria. It appeared that the only way to do this was via sea, and therefore the Russians were hiring as many merchant ships as possible to help with the mammoth task.

Seeing an opportunity, Shackleton and a consortium that consisted of Thomas Garlick, an accountant, and George Petrides, the brother of an old schoolfriend, looked to provide shipping to help with evacuating 40,000 Russians from the port of Vladivostok. The prospective deal promised to pay between £12 and £40 a head, depending on rank. Shackleton excitedly wrote to Emily, 'There is a chance of our little steamboat company doing a big deal in a few days. It would mean £10,000 to me, but I cannot go into details now. It is awfully exciting.' He did not, however, realize that he was only a minor pawn in a much bigger game.

At the time, the Russians were negotiating a far larger contract with a major German-American shipping line and were trying to knock the price down. With the likes of Shackleton accepting lower fees, it gave the Russians some leverage in securing the contract they desired with the shipping line. Shackleton was once more left high and dry.

Following yet another bungled business deal, and in desperate need of money for his young family, Shackleton again turned to William Beardmore. While they had previously spoken about Beardmore bankrolling an expedition, now Shackleton just needed a job. It was quite the fall from grace. Thankfully, with Beardmore having taken a liking to Shackleton, he found him a role at his Parkhead Works in Glasgow. There, he would serve as the secretary of a small committee set up to assess the design of a new gas turbine engine. In effect, he was to be a glorified note-taker, at a salary of just £30 a month.

He tried to convince himself, and Emily, that this was once again just a stepping stone to greater things, writing, 'I may become a director before long. If I had say ten thousand pounds in the business it would pay from 10% to 14% and then the directorship is at least worth one thousand a year so we ought to do well.'

Once more, this was not a role that suited Shackleton. He should have known by now that his rambunctious personality was ill suited to life behind a desk. This became very clear to his superiors, who found that his minute-taking left a lot to be desired. More suitable work was quickly found in the hope of making the most of the skill Shackleton possessed in abundance: entertaining. Shackleton thus became an ambassador for the company, entertaining clients at Parkhead or in London. Naturally, they warmed to the cheery Irishman, whose thrusting style and way with words frequently had admirers in the palm of his hand. Beardmore's secretary, A. B. MacDuff, remembered Shackleton fondly from his time at Parkhead, 'Even if I was in the thick of things, I'd give them up to do what he wanted, I'd such a liking for the fellow.'

For a man who seemingly desired to make his fortune, it was said that he regularly forgot to pick up his wages. 'He left the salary with us and forgot all about it for five months,' MacDuff remembered. This illustrates that, to Shackleton, money was not his primary motivator. He was lucky in that he had Emily's money to fall back on, but more than anything, he wanted to be the hero, to mark his name in history and be remembered. If money was a by-product of that, then all the better, but that wasn't what would keep him up at night and fill his head with dreams. Every job he took seemed to be more about prestige and proving himself than anything else.

Increasingly restless, and tiring of the regular commute, Shackleton once more began to look into the prospect of mounting an expedition, maybe this time all the way to the South Pole itself. In a letter to Mill, he wrote, 'What would I not give to be out there again doing the job, and this time really on the road to the Pole!'

For now, at least, he kept his ambitions a secret from Emily. He knew such a fancy might cause some distress when he had only just settled into regular work. The money he earned, when he did bother to pick up his salary, was also being spread thin. His son, Raymond, was growing fast, while his father had had to stop treating heart patients because of his increasing deafness, so Shackleton had to send money to London to help out.

Nevertheless, Shackleton remained determined to mount another expedition come what may, particularly when in October 1906 Roald Amundsen roared into San Francisco having completed the first navigation of the North-West Passage. Soon after, the American Robert Peary set a new furthest north record, getting to within 200 miles of the North Pole. All of this was like a red rag to the restless Irish bull named Shackleton. At this he became ever more intent on returning south and escaping his day-to-day drudgery.

It soon became clear to many that Shackleton's mind was now elsewhere. On one of his regular visits to chat with Beardmore's secretary, MacDuff asked if he was considering another adventure. 'Yes,' he replied. 'I want to go on a further expedition soon. This time, I want to command it myself.'

When Shackleton told Elspeth Beardmore of his latest

plans, he was delighted to find that she was very supportive and felt that she might even be able to persuade her husband to fund it. Everything looked to be gathering pace when Shackleton was once more stopped in his tracks: Emily was pregnant again. It seemed he would forever be chained to his desk, dreaming of better days, while others conquered unknown worlds.

15

The famed explorer Robert Peary once said, 'The lure of the ice, it is a strange and powerful thing.' I couldn't agree more. No matter all the discomfort and failure you might have endured, the urge to return is mesmeric. I've often described it as having the same pull as the one a smoker feels when trying to shun thoughts of tobacco. The memories of gangrene, crotch rot and frostbite (I lost the ends of five fingers) are eclipsed by the rose-tinted spectacles through which the prospect of a grand adventure is viewed. As another American once said, 'Nothing is more responsible for the good old days than a bad memory.'

This was certainly the case for Shackleton. No matter where he found himself, he could never be rid of the fanatical voice reverberating in his head urging him to return. The Danish even have a word for this – *polarhullar* – which translates as an ache for the polar regions. Most of all, he wanted to return to satisfy his ambitions, etch his name in history and make his fortune, as well as prove Scott wrong. But even he was aware that such a trip was fanciful with Emily pregnant again and finances already stretched.

Just before Christmas 1906, Emily gave birth to their daughter, Cecily. Despite all of his dreams apparently being dashed, Shackleton was delighted. 'You will be pleased to hear that the most interesting Christmas present I got was from my wife,' he wrote to Mill, 'who on Sunday morning introduced to the family a splendid little girl.' And yet while he tried to play the doting father, at the same time, without breathing a word to Emily, he was still somehow trying to work out how he could return south.

Sketching out a rough plan, he decided that he would base his camp at Hut Point, just as the *Discovery* had done. From there, he would attempt to reach the South Pole, travelling through the mountains which he, Scott and Wilson had identified as a potential gateway to the south.

Shackleton also continued to share his plans with Elspeth Beardmore. She was a good listener, prepared to indulge his heroic fantasies, and she always sought to encourage him. After all, she had nothing to lose in doing so and, being childless herself, she could not truly imagine Emily's position. 'You are always so cheerful,' he wrote to her, 'and make me feel so much better after I have seen you.'

Over the years, this relationship has certainly set plenty of tongues wagging. While there is no definite evidence that Shackleton and Elspeth enjoyed a romantic affair, their letters certainly hint at their being more than just friends. 'Elspeth . . . you have always been such a real friend and confidant to me,' he wrote, 'that it is to you alone I can talk . . . You looked so beautiful the other night.'

Such a relationship is reminiscent of Shackleton's early pursuit of Emily. He always seemed more interested in the chase rather than the final destination. This was of course the case in both his work and his romantic life. As he later wrote while down south, 'It must be part of my life that I go on striving for things that are out of reach.' The wife of one of Scotland's richest men was definitely that, and he was no doubt flattered that she in turn expressed such an interest in him. Whether there was any more to it than that we can but speculate, but the two certainly spent a lot of time together and shared their innermost secrets.

After weeks of talks with Elspeth, who continued to build

his confidence, Shackleton decided to break the news to Emily that he planned to return to the Antarctic. 'I shall come back with honour and with money and never part from you again,' he promised. Emily might have been forgiven for putting her foot down. They had little money, and she had just given birth to their second child. Surely it was time for the man she called a 'boy' to grow up? But Emily understood his restless spirit and his need to prove himself. She knew that no matter what she said there would be no stopping her ambitious husband. 'She was never the woman who wanted to raise a finger to make it difficult for him to go,' her daughter later recalled. Emily saw her place as looking after the family while her husband sought to conquer the south. However, it was all very well having Emily's approval and Elspeth's encouragement, but to make the dream a reality Shackleton now needed cold, hard cash.

In early 1906, Shackleton prepared a proposed budget in a formal document headed with the meaty title 'Plans for an Antarctic Expedition to proceed to the Ross Quadrant of the Antarctic with a view to reaching the Geographical South Pole and the South Magnetic Pole'.

Vessel fully equipped	....	£7,000
Provisions for three years – Clothing, Salaries for staff, Scientific equipment, Oil, Sledge equipment including dogs and ponies, Carriage of dogs, ponies and passages, General Expenses	....	£10,000
Total	....	£17,000

The figure for the total later proved to be wildly optimistic; Shackleton would need at least £50,000 to fund the expedition. Unrealistic optimism was, however, very much the name of the game. As the Scottish philosopher William H. Murray once said, 'The moment one commits oneself, then Providence moves too.' Shackleton was a great believer in providence, and the word features many times in his diaries. He therefore threw himself into the venture, but many of the hardened businessmen whom Shackleton initially approached for sponsorship saw through him, and his plans.

These men, who had made their fortunes counting every bean, could smell a dreamer like Shackleton a mile off. To most, Shackleton had no great Royal Navy background, as Scott had, and had only ever been on one expedition, which the public now knew, thanks to Scott's book, had ended with his being pulled on a sledge as an invalid and sent home early. On the face of it, Shackleton hardly inspired the confidence they needed to believe that he was worth throwing money at. For daring to doubt him, Shackleton noted some seventy 'Negatives' on a blacklist, vowing to prove each and every one of them wrong. I found years later that to gain donations it is necessary to think of yourself not as an explorer but as a salesman. In time, I acquired a more direct patter, and the same was true for Shackleton, who became one of the era's arch-exponents of the art of sponsor-getting.

As Shackleton presented this so-called 'opportunity' to friends and family, his cousin from Ireland, William Bell, promised to chip in, as did another family connection, Emily's older brother, Herbert Dorman. Mining ever deeper into those who had deep pockets and with whom he had some sort of connection, Shackleton reacquainted himself with Gerald Lysaght, the steel manufacturer he had first met many years before on the *Tantallon Castle* liner. While he managed to squeeze only a token donation from Lysaght, every penny would count. His greatest success came when he recalled the £1,000 cheque the elderly spinster Elizabeth Dawson-Lambton had donated towards the

Discovery. Reintroducing himself, he found she remembered him well and she wrote him another cheque for £1,000. However, this was all just a drop in the ocean towards the minimum £17,000 he required for ship, supplies and personnel.

What Shackleton really needed, rather than chasing lots of smaller sponsors, was a benefactor who believed in him, like Llewellyn Longstaff, who had effectively bankrolled the *Discovery* expedition himself. Knowing Elspeth Beardmore was a great supporter of his, and also knowing her wealthy husband well by this stage, Shackleton now turned all of his attentions to him. It was make or break. 'I at last took my courage in both hands and asked him straight out,' he remembered.

Beardmore liked Shackleton but, nevertheless, he had some – understandable – reservations. He certainly would not be throwing money at an unproven investment like Shackleton without limiting his exposure. As such, he merely agreed to guarantee a £7,000 loan with Clydesdale Bank, which Shackleton would pay off with the 'first profits' of the expedition. While this was not what Shackleton had envisaged, he hoped that when other potential sponsors saw that a wealthy individual such as Beardmore had invested in him at least a little, then they might be more reassured and open their cheque books in turn. That was the idea, anyway.

When Ginny and I set out to raise money for our expeditions we always vowed not to borrow a penny. Every single item had to be sponsored, from the boat to the butter. We certainly did not want any debt hanging over us, so much so that we didn't open a bank account or have any credit facility. Of course, asking for things for free took a little longer, seven years, as I mentioned earlier, to raise over £30 million in sponsorship and donations for the Trans Globe Expedition, but we could at least then focus on the job in hand, rather than fretting about paying back an enormous debt on our return. For now, Shackleton was just desperate to get back to the ice, no matter what the consequences.

While he had still not met the minimum mark of £17,000, with Beardmore's guarantee he now felt that he at least had enough money to get things rolling. In early 1907, Shackleton moved his family to London and rented a small office in Regent Street, which he crammed with interesting polar gear to impress visitors, whom he would then ask for a donation. But in paying rent for an office, Shackleton was cutting further into his ever-dwindling finances.

From his costly new London base, Shackleton now targeted one sponsor he was very confident would come on board: the Royal Geographical Society. He knew that this was crucial. Any backing by the RGS would lead to backing from the King, which would then surely open the floodgates. With Roald Amundsen scheduled to give a talk at the RGS on 11 February 1907, in which he would discuss his incredible success in the North-West Passage, Shackleton decided this was an opportunity too good to miss. It would allow him to hear the great man speak, while also being an opportunity to bang the drum for his own expedition, now named the British Antarctic Expedition. 'I will try and get a few big fish [life members] into my net and try others,' he told Mill. 'Am hoping that by the end of the week I will have all the money guaranteed and shall announce it before the 12th. I think the end is in sight.' Yet other explorers had the same idea as Shackleton, and competition for sponsorship was fierce.

One such rival was Henryk Arctowski, a Polish scientist who had survived the *Belgica* expedition. He also planned to be present to unveil his Belgian expedition to reach the South Pole with motorized sledges. French explorer Dr Jean-Baptiste Charcot was also circling, as were Barne and William

Speirs Bruce, who had inspired Shackleton to lead his own expedition in the first place, all looking to raise money for their own expeditions.

When Arctowski bumped into Shackleton at the Royal Geographical Society and informed him of his intention to announce his South Pole plan that very night at the Society's dinner, Shackleton reacted like a scalded cat. Knowing he had to move first, and fast, he announced his own plans minutes before Arctowski could do likewise. After the dinner, he then quickly contacted *The Times* to describe his project as a national venture, one that would beat off a foreign rival.

The Times duly reported, 'As similar expeditions are being organized by other nations, what may be called an international attack on the south polar regions will be made which will be followed with keen interest.' While the announcement served as a warning to any other contenders who might be planning to operate on 'his' zone of coastal Antarctica, Shackleton knew that such a public announcement would also gain him British support. After all, it should surely be a British subject who had the honour to be the first man to reach the South Pole.

Using this nationalistic spin to his advantage, Shackleton warned the RGS that, with rival foreign expeditions already making headway, there was no time to waste. He needed their support, and their money, quickly. However, the Society was not convinced.

Keltie, the Society's Secretary, and Sir George Goldie, who had succeeded Markham as President, greeted Shackleton's plans with indifference. In particular, they felt his plans still had 'many mysterious questions and hints' rather than hard answers. That might very well have been the case, but what they failed to tell Shackleton was that the RGS was already planning to sponsor Scott's plan to return south, and believed he was the better bet.

Unaware of the duplicity of the RGS, or how far advanced Scott's proposed plans were, Shackleton continued to plan ahead regardless, now turning his attention to the crew. Naturally, Wilson, his close friend from the *Discovery*, was his first choice to be his right-hand man. 'I want the job done and you are the best man in the world for it,' Shackleton wrote to him. 'If I am not fit enough to do the southern journey there could be no one better than you.' However, Wilson had recently taken a job on the Scottish moors, investigating an epidemic that was wiping out grouse. 'I am honour bound to carry this grouse work through,' he told a shocked Shackleton, just two days after receiving his letter. Knowing that Shackleton would continue to chase him, he signed off with 'don't waste more money on long telegrams.'

Not securing Wilson was a blow, and Shackleton soon found that others from the *Discovery* were also unavailable: Armitage, Hodgson, Skelton and Barne. The true reason for this soon became apparent. Upon approaching George Mulock, the man who had replaced Shackleton at McMurdo Sound in 1903, Mulock revealed that he had already committed to join Scott. Suddenly, the reasons behind his rejection by the *Discovery* crew, as well as by the RGS, became clear.

What really hurt Shackleton now was that while Scott had been busy recruiting the crew from the *Discovery*, Shackleton himself had not been asked to return. This was a shattering dent to his confidence. Suddenly, he felt very angry and very much alone.

Meanwhile, with Scott having returned to the Navy for the time being, the news of Shackleton's proposed expedition reached him while he was at sea. The timing could hardly have been worse. He had recently taken command of the battleship *Albemarle* and was practising dangerous, and highly skilled battle tactics, in the Atlantic, west of Portugal.

On a pitch-dark night, as eight multi-ton steel warships sped through the darkness, without lights, Scott left the bridge to signal a message. Moments later, he felt the horrific impact of *Albemarle*'s bows striking another ship.

News of the accident, linked with Scott's name, made front-page news. An official inquiry subsequently cleared him of all blame, but on the day that Scott first heard of Shackleton's polar intentions he was still under severe stress, fearing a court martial and disgrace. Now, with the prospect of the man he had called an 'invalid' threatening his goal to reach the Pole, he was incandescent. In the circumstances, his response to Shackleton could only be described as pretty restrained. On 18 February 1907 Scott wrote:

> *My Dear Shackleton*
>
> *I see by the* Times *of Feb. 12th that you are organising an expedition to go on our old tracks; and this is the first I have heard of it. The situation is awkward for me as I have already announced my intention to try again in the old place, and have been in treaty concerning the matter. As a matter of fact I have always intended to try again but as I am dependent on the Navy I was forced to reinstate myself and get some experience before I again ask for leave, meanwhile I thought it best to keep my plans dark – but I have already commenced fresh preparations and Michael Barne is in London seeing to things in preparation for August when I shall be free to begin work myself.*

Letters between the two men followed one another thick and fast. Both would have known that the other's main goal was to be first to reach the South Pole, no matter how many other scientific and subsidiary trips were discussed and dissembled. Realizing that Shackleton did not intend to stand down, Scott tried to make things as awkward as possible for him.

When he learned of Shackleton's plans to land at McMurdo Sound, Scott fired off another letter, telling him such a destination was off limits. 'I think anyone who has had anything to do with exploration will regard [McMurdo Sound] as mine,' Scott wrote. 'It must be clear to you now that you have placed yourself in the way of my life's work. If you go to McMurdo Sound you go to winter quarters which are clearly mine.' He signed off with a reminder to Shackleton of all he apparently owed him. 'I do not like to remind you,' he wrote, doing just that, 'that it was I who took you to the South or of the loyalty with which we all stuck to one another or of incidents of our voyage or of my readiness to do you justice on our return.'

Scott clearly could not enforce such a claim to McMurdo Sound, even if he had been the first to overwinter there. Indeed, it seemed not to matter to Scott that Borchgrevink had discovered McMurdo Sound two years before the *Discovery* expedition and he had put no barriers up to Scott using it. As Mill told Shackleton, any territory was 'absolutely open and free to anyone who has the courage and perseverance and good luck to reach it'.

McMurdo Sound being off limits would seriously dent Shackleton's plans, in more ways than one. Not only was it well known to him, and the best-known landing spot to boot, Shackleton had also made a strong point in his approach to Beardmore, and other sponsors, that his knowledge of the area gave his plan a high chance of success. This was one of the few aces he had up his sleeve in comparison to other potential expeditions. If he now had to tell his current and prospective sponsors that some of this vital experience would not be of use, there was a possibility that they would pull out.

Shackleton was, however, also aware that a dim view would

be taken, and much of the glory tarnished, if he were now to set foot on McMurdo Sound, no matter how difficult this made his expedition. In Edwardian England, much was made of being seen to be fair and gentlemanly. In that respect, Shackleton was aware that Scott was viewed as the Royal Geographical Society's golden boy and that, therefore, he had to tread a fine line if he were not to tarnish his own name.

Nevertheless, the underlying feud between Scott and Shackleton now looked set to erupt into the open. 'Shackleton owes everything to me,' Scott raged to Keltie. 'I hold it could not have been playing the game for anyone to propose his expedition to McMurdo Sound until he had ascertained that I had given up the idea of going again.' Now looking beyond the issue of McMurdo Sound, Scott tried to stop Shackleton's expedition in any way he could by claiming that his venture might 'ruin the cause of true exploration'. He also hoped to get others more powerful than himself to step in, stating, 'It is a question of how far the Society should condone an act of disloyalty.'

Many of Scott's outbursts at this time must be taken in context. He was not only under severe stress following the incident at sea, he was also struggling to provide for his elderly mother. He might have also legitimately believed that an expedition led by a man such as Shackleton, who had never led an expedition before, and had returned an invalid on the last trip, might prove to be such a disaster that it would halt any further expeditions for years to come. This had certainly been the case following the disastrous 1845 Franklin expedition. If something similar should occur with Shackleton's expedition, then Scott's own plans could be dashed. Keltie stirred the pot by writing to Scott that Shackleton was 'not absolutely sound', and 'Heaven knows what may happen if he starts on his journey Pole-wards.'

While this might explain part of his hostility to Shackleton's plans, Scott was so enraged that he could not resist bringing up the past. 'I personally never expect much in this sort of work from a man who isn't straight,' he told Keltie. 'Shackleton is the least experienced of our travellers and he was never very thorough in anything – one has but to consider his subsequent history to see that he has stuck to nothing and you know better than I the continual schemes which he has fathered.' These scathing statements are contrary to everything Scott wrote and said at the time of the *Discovery*.

Such was Scott's mood that he even wrote to Edward Stanford, the owner of Stanfords Bookshop, to reveal his anger at a map of Antarctica which the shop had produced to identify the exact 'furthest south' location 'reached by Scott and Shackleton'. Scott pointed out that marking that historic geographical location with both his name *and* Shackleton's clearly 'implied dual leadership' . . . which was 'not in accordance with fact'. Stanfords quickly reassured Scott that 'everybody knew that you were the leader' and duly removed Shackleton's name from the map.

Matters now threatened to turn ugly. To try to mend relations Shackleton hoped that Wilson, a mutual friend, might be able to work out a compromise. But, to his surprise, Wilson chose to back Scott. 'I do wholly agree with the right lying with Scott to use that base before anyone else,' he wrote. 'I think you ought to offer to retire from McMurdo Sound as a base.' Ominously, he also warned that if Shackleton did decide to winter at McMurdo Sound, then 'the tarnished honour of getting to the Pole even as things have turned out will be worth infinitely less than the honour of dealing generously with Scott.'

Now that Wilson had shown where his loyalties lay,

Shackleton turned to Markham to mediate. Yet Markham not only chose Scott's side but also proceeded to rubbish Shackleton. 'He [Shackleton] has behaved shamefully,' Markham told Scott. 'It grieves me more than I can say that an expedition [*Discovery*] which worked with such harmony throughout should have had a black sheep.'

Shackleton knew that, in the public eye at least, he was fighting a losing battle. Scott was the Navy man and a national treasure. No matter how right Shackleton might have been, he realized that to persevere would do him no good, and was also costing vital time. If Scott's goal was to stall Shackleton, then it was certainly working. As such, Shackleton now changed tack.

In March 1907, with Scott having returned from the Atlantic, the two men met in London to thrash out an agreement. By May, they had agreed that Shackleton should have a free rein anywhere east of 170 degrees west. Agreeing not to step foot on McMurdo Sound, Shackleton told Barne, 'I would rather lose the chance of making a record, than do anything that might not be quite right.' This was extremely generous of Shackleton. While the agreement has to be viewed through the prism of time, there is still a case to be made that this was an unfair and unjust request on Scott's part.

Shackleton had so far somehow managed to keep his composure and cheery good nature, but then Wilson pushed him too far. Seeking an alternative base to McMurdo Sound, Shackleton considered landing at King Edward VII Land. This was in part thanks to Armitage, who had proposed the base to Scott while on the *Discovery*, believing that it promised an easier route to the pole. However, when Wilson heard of this, he wrote, 'Don't on any account make up your new plans until you know definitely from Scott what limits he puts to his rights.'

This was too much for Shackleton to bear. He had done all he could to acquiesce to Scott, even though he was under no obligation to him, but this effectively put his expedition into Scott's hands. He was also dismayed that Wilson had chosen to get involved to such an extent that their friendship was now under threat. Determined not to shift another inch, he wrote, 'I do not agree with you, Billy, about holding up my plans until I know what Scott considers his rights. There is no doubt in my mind that his rights end at the base he asked for, or within reasonable distance of that base. I consider I have reached my limit and can go no further.'

Matters were now reaching boiling point, with both men desperately trying to raise funds from limited sponsors. Scott seemed to be holding the upper hand, by virtue of his Royal Navy background, successful *Discovery* expedition and the Royal Geographical Society seal of approval. Keltie did not even try and hide his loyalty to Scott, writing 'you need not have any doubt that if the two expeditions are to set out on which side the heartiest wishes for success will be'. On the face of it, Shackleton's prospects did not look encouraging. However, he still had some things in his favour. Scott was tied to the Navy for a further two years, while Shackleton was free to continue with his plans. Between the two, Scott was the practical choice, but Shackleton, the wild card, armed with an optimistic, cheerful and determined personality which made people forget his shortcomings and open up their cheque books, wasn't out of the race just yet.

Having persuaded all of his financial sponsors that his change of camp was in no way negative (this would have been extremely difficult, I suspect, had any of them actually known anything about Antarctica), he aimed to set off in August 1907. He knew that if he failed to meet this target, a very uncertain future lay in store.

Still requiring significant investment, Shackleton commenced hunting for the most expensive, and crucial, item on his shopping list – his ship. However, at this stage, most suitable ships were out of his reach. When a modern ship with new engines and room for fifty men, ideal for the journey south, was presented to him, Shackleton baulked at the £11,000 price. That was more than all the money he had in the bank, before he had even paid for further equipment, provisions and crew.

Short of options, Shackleton was subsequently presented with an older wooden sealer called the *Nimrod*. Built in Dundee in 1866, she could move at just 6 knots, and was only 136 feet in length. She had spent almost fifty years at sea, and Shackleton commented upon seeing her, 'She was much dilapidated and smelt strongly of seal-oil.' If she was to be sea ready, the ageing and creaking *Nimrod* would clearly require extensive repair work and modifications to withstand the rigours of the Antarctic, but at a knock-down price of £5,000, she was all Shackleton could afford.

However, in June 1907, with the *Nimrod* in London for a comprehensive refit, which included a new engine, Shackleton ran out of money and, with it, hope. Unless something miraculous happened, he faced ever more shame, frustration and humiliation, with Scott having all the glory to himself.

16

If Shackleton was to avoid scrapping the expedition, he desperately required a further £8,000. Yet if he was still intent on leaving in August, and beating Scott, then he could not halt the refit of the ship. Keeping his cards close to his chest, he told the refitters to continue with their work, all the while hoping that he could somehow plug this financial black hole.

By now, he had approached any man or woman of any wealth with whom he had any sort of connection. After a flurry of rejections, it seemed he was out of options. Spreading his net, and hopes, ever wider, he now scoured Britain for anyone who had a record of sponsorship of any kind. Finally he came across Edward Cecil Guinness, the Earl of Iveagh.

The Earl was the head of the Guinness brewing dynasty, and had donated over £1 million to various charitable causes over the years. However, these usually encompassed worthy causes such as housing and slum clearance. As of yet, the Earl had shown no interest in sponsoring anything like an expedition. He had also never met Shackleton, but Shackleton planned to lean on two things they did have in common: they were both Irish, and Freemasons, with Shackleton having joined the order upon returning from the *Discovery*.

Shackleton knew that to stand any chance of success he could not appear to be desperate. Therefore, with charm, bluff and bluster, Shackleton made Iveagh feel as if it would be to his benefit if his name was linked to such a historic expedition. The purchase of the *Nimrod* might also have helped; a ship already acquired, and undergoing a refit made the expedition appear as if it was on track, when it was anything but.

The Earl was, however, far too shrewd to be totally taken in by Shackleton. Rather than sponsor him, he agreed to guarantee a loan of £2,000, on the proviso that other sponsors could be found to raise the total to the much needed £8,000. This was far from what Shackleton required, but it was better than nothing. Somehow, he still had to raise £6,000, with time running out.

Meanwhile, he still had to recruit the crew, with most of those suitable having been taken by Scott. He was essentially looking for the best of the rest, hoping to find some unpolished gems. From his cramped Regent Street desk, he scoured through over 400 applications, most from men nowhere near suited to such an excursion. Working late into the night, he was barely seen at the family home in Bayswater. Revealing the stress he was under, he wrote to Elspeth Beardmore, 'There are 1000 and 1 things I must do and the time seems all too short to do them in.'

It is a wonder how Shackleton kept his head. Money was running out, all suitable crew were apparently taken, an expensive ship refit was underway, and most at the Royal Geographical Society and from the world of expeditions had turned their back on him. Common sense would have told any lesser man to quit, but Shackleton seemed to almost will the expedition to life.

Trying to find suitable crew was proving to be a challenge, but he remembered meeting Jameson Boyd Adams in 1905 while he and Emily were taking a holiday in Scotland. He knew that Adams was in the Merchant Navy and was also a Royal Navy Reserve. While he had no specific expedition experience, Shackleton felt he would be a good fit. Having persuaded him to forgo a career in the Navy, for now at least,

Adams was appointed as Shackleton's second-in-command, and subsequently helped interview some of the more suitable applicants, albeit with certain restrictions.

Knowing from bitter experience just how vital fitness and health were to an expedition, Shackleton decreed that any candidate 'must be free from any heart troubles'. He also wanted to ensure that the crew would be classless, so that there would be no divide, as there had often been on the *Discovery* between the Royal Navy and merchant seamen. 'The temperament of the various members of the expedition is one of the most serious and important factors in such a case as ours,' he wrote. 'I feel that the success of our work depends as much on the general attitude of the members to each other as on the work they individually have to do.'

Again, in an already weakened field, these provisos seemed only to further whittle down the choice of crew. Shackleton was, however, able to persuade two old *Discovery* colleagues to come on board: Frank Wild and Ernest Joyce. It was only through sheer luck that Shackleton managed to recruit Joyce. Having spotted him on the top of a London bus, Shackleton waved him down and asked him to join the expedition. Joyce, who was then in the Navy, was only too happy to accept, on the proviso that he be part of the pole team and also receive financial compensation from Shackleton for the loss of his Navy pension. Shackleton agreed, adding another credit to the long list to be paid on the expedition's return.

While most of the old *Discovery* crew remained off limits, Shackleton had met a handful of men over the course of his career he felt he could rely upon. Adams was one such individual, but there was also twenty-eight-year-old Dr Eric Marshall, whom Shackleton had met on the social scene. Marshall was a prized rugby player, and Shackleton had been impressed with his intelligence and athletic physique. George

Marston, who was friends with Shackleton's sisters, and was described as the 'fat comedian with the permanent smile', also found his way on board as chief illustrator.

Lieutenant Rupert England, who was assigned to captain the ship, was known to Shackleton, having been chief officer on the *Morning*, the relief ship that had been sent to assist the *Discovery*. While he had never captained a ship before, he was vouched for by William Colbeck, the captain of the *Morning*, whom Shackleton had originally eyed for the role. With this, and with England having at least previously been to the Antarctic, Shackleton saw no reason not to appoint him to the position, while twenty-three-year-old John King Davis would serve as first officer.

More than any personal connection, Shackleton relied upon gut instinct when recruiting the crew. This is something I have also relied upon when sizing up potential expedition members. Initially, I put most prospective candidates through their paces on the mountains of North Wales, looking for professional, dogged people, usually from the SAS Territorials. But over time I came to find that a positive personality and being able to work as part of a team were just as vital as being able to endure harsh conditions. Ultimately, I found it better to search for conventional people with everyday jobs who score highly under three criteria: level-headedness, patience and good nature towards others.

With this in mind, for the Trans Globe Expedition I eventually recruited Charlie Burton, a retired corporal from the Royal Sussex Regiment and by then a security officer; Oliver Shepard, ex-Welsh Guards and at the time a Whitbread beer salesman; and Geoff Newman, a print expert. None had expedition experience, but all seemed suitable personality-wise.

Shackleton's interview style was telling in this respect. He was clearly looking for certain types of character who he felt could be of value during a long expedition. Upon meeting twenty-one-year-old Raymond Priestley, he asked several random, seemingly unrelated questions of the young geologist, such as if he could sing. 'He must have asked me other questions but I remember those because they were bizarre,' Priestley later recalled. Priestley was young and untested, but Shackleton liked what he saw. It is a testament to his ability to see the potential in men that Priestley would go on to have a stellar career, becoming a distinguished geologist, receiving a knighthood and eventually serving as chairman of the British Association for the Advancement of Science and even as President of the Royal Geographical Society.

In recruiting the crew, Shackleton came to realize an angle he had not yet played. As of yet, the scientific element of the expedition had been an afterthought for him; he was far more interested in the glory of reaching the pole. Now he saw that including a scientific element could help gain official approval and even sponsorship.

I also learned this the hard way. The media, and particularly key sponsors, give significant brownie points for expeditions with solid scientific research programmes rather than those which are just macho stunts. I've therefore always ensured that my polar expeditions have a significant scientific element, whether it be meteorology, high-frequency propagation, cardiology or blood analysis, as often these pay for the bulk of the expedition costs.

With this principle in mind, Shackleton sought out the best scientists he could find, recruiting the experienced Scottish biologist James Murray as well as Dr Alistair Forbes Mackay. These men might have made the expedition appear a serious scientific endeavour, but it was a twenty-year-old assistant geologist who would prove to be a great help to Shackleton's sponsorship problem.

Sir Philip Lee Brocklehurst was not only a budding geologist, he was also an Etonian baronet from the manorial Swythamley Hall in Staffordshire. After being introduced to the young baron, Shackleton swiftly saw that he not only had the necessary qualifications to serve on board as a scientist but he also had vast wealth and connections. To access this, all he had to do was persuade the young baron to join up, which he did, thanks to Shackleton's considerable powers of persuasion.

Following this, Shackleton wasted no time in arranging to meet the baronet's widowed mother, Lady Annie Lee Brocklehurst. Charmed and flattered by the good-natured Irishman, she not only agreed to guarantee a loan of £2,000 but also also did something far more important: she flung open the doors to high society. Soon, thanks to these contacts, Shackleton was able to fill much of the £8,000 financial black hole (although this was now growing bigger). And there was still better news to come.

Major-General John Fielden Brocklehurst, a cousin of Lady Brocklehurst, was a renowned soldier and equerry to Queen Alexandra. Shackleton realized that this connection could provide him with a route to the King himself, bypassing the Royal Geographical Society. The royal stamp of approval would be his passport to further sponsorship, media publicity and success.

Shackleton immediately wrote to the King's private secretary to ask for the monarch's support, dropping in the Brocklehurst name. While he did not get the royal patronage he sought, the King agreed that, should the *Nimrod* attend the annual Cowes Week regatta, just as the *Discovery* had, he would be happy to come on board and give the expedition a royal send-off. Having been in such dire straits just weeks before, this gave Shackleton, and the expedition, an instant lift.

But, as always seemed to be the case, bad news swiftly followed good.

Just as it looked like Shackleton was getting the finances in better shape and gaining a royal send-off, there came the threat of family disgrace. In July 1907, the Irish crown jewels, valued at £40,000, were stolen from a safe in Dublin Castle. No doors or locks were broken, while the safe was opened with a key. It appeared to have been an inside job, and all eyes were on someone known to Shackleton.

The safe had been kept in the office of Sir Arthur Vicars, whom Shackleton's brother, Frank, had worked for, as one of the heralds at Dublin Castle. It was already known that Frank was in deep financial trouble and under investigation for fraudulent trading on the stock market. Unsurprisingly, with the motive and means, Frank was the primary suspect.

Making matters worse, Frank was a homosexual, a criminal offence at the time, and was known to take part in orgies organized by the Duke of Argyll, the King's brother-in-law. Such a revelation would have been explosive, not just for Shackleton but also for the royal family. The press would have a field day linking the story to the explorer and the King.

While Frank had not yet been arrested, and would eventually be exonerated by a 'sham' committee who sought to protect guilty parties in high places, his name was being rubbished as a debtor. That alone could have been ruinous for Shackleton, who was still desperately trying to raise funds. Having learned that Frank required £1,000 to avoid disgrace, Shackleton begged Beardmore for a short-term loan, to be repaid before the *Nimrod* set sail. Beardmore came good, but Shackleton did not tell him what the loan was for, although it seems he did tell Elspeth. 'Frank has caused me a lot of worry and expense,' he told her. 'But now he is out of his trouble and will pay me back the money he owes.'

For now, at least, Shackleton had avoided disgrace and could turn his attention back to purchasing the equipment and provisions needed for the expedition. And this time there was someone who was at last willing to help. Fridtjof Nansen, the world-famous Norwegian polar explorer, was now serving as the Minister to the Court of St James in London on behalf of the newly independent Norway. Nansen had achieved greatness in the frozen north and now dreamt of conquering the south. This did not, however, stop him from giving helpful advice to British South Pole wannabes, including Shackleton and Scott. After bouncing around several ideas with Nansen, Shackleton soon had a better idea of what to look for, basing his decisions on his own experiences, plus Nansen's advice.

Since Shackleton had suffered from scurvy on the march south, he was as aware as anyone of the importance of a good diet. He therefore chose all his food supplies with an eye on antiscorbutic properties and was amply sponsored by the likes of Rowntree, Colman, Glaxo and Lipton.

In addition to standard staple foods, Shackleton, who personally loved tasty dishes, packed copious luxuries, including sweet jams, exotic soups, ginger biscuits and various chicken and mushroom goodies. However, he also made sure there would be plenty of pemmican for making hoosh, knowing it was vital to provide energy and warmth when on the march. William Roberts, an internationally employed hotel chef, was also engaged as cook, to make the dishes as sumptuous as possible.

Shackleton also sought to take advantage of recent advances in technology. A company called Venesta provided lightweight waterproof cases made of composite board which could not only be used to pack things in on the boat but could then be remodelled as partitions and furniture in the expedition hut. Using them would also save more than 4 tons in weight. In addition, the hut itself, made of pine timber, with insulated roofing, felt and granulated cork packaging, was prefabricated, making it far easier to store than the one taken on the *Discovery*.

Shackleton also had another purpose in mind for the Venesta boards. With the *South Polar Times* having proven very popular on the *Discovery*, he had managed to obtain a printing press free of charge, as well as ink, paper and type, courtesy of Sir Joseph Causton and Sons, the well-known City publishing firm. The multipurpose boards would, he thought, be the perfect material for the covers.

Other modern gadgets that Shackleton took south included the latest photographic gear for still and movie action coverage. He hoped these images would in time become extremely valuable, especially with the lucrative book deal and lecture tour he envisaged on his return.

Travelling to Norway, Shackleton was keen to take advantage of all the cutting-edge Norwegian developments in expedition equipment. Nansen's advice proved invaluable in this. Shackleton duly ordered thirty sledges, along with fifteen reindeer-skin sleeping bags, plus reindeer-leg-fur finnesko boots from Lapland as well as wolf- and dog-skin mitts. Although not particularly a ski enthusiast, he also ordered plenty of skis, boots and sticks from Oslo.

Admirable as it is that Shackleton planned to take with him the latest equipment, he would find out if it would be of any benefit only once he was in the Antarctic. If it proved useless, it would be too late to do anything about it. Much as I found my training in Greenland and North Canada helpful in readying myself for Antarctic and Arctic conditions, it also allowed me to test out various items to see what worked and what didn't. With more patience, and perhaps foresight,

Shackleton could have done exactly this, and in the process saved time and money, while also helping to ensure the expedition's safety.

One of his more crucial decisions was which animal would be the most useful in helping to haul provisions and equipment to the pole. On the *Discovery* expedition, Scott had used a mixture of dogs and man-hauling. The dogs had, of course, turned out to be troublesome, but Shackleton did say after the expedition, 'We only had twenty-three dogs when we started. I wish we had had about sixty or seventy, for then we would have reached the Pole.' However, in the years that followed, it seems Shackleton came to believe they were more trouble than they were worth. 'Dogs had not proved satisfactory on the Barrier surface,' he later wrote. He now considered alternatives, one of which was ponies, although they had never before been used in the Antarctic, so he could not know if they would be of any use.

Quite who converted him into this line of thinking is not traceable. Nansen had extolled the use of dogs to haul sledges, and skis with sticks. It is therefore most likely that acquaintances from the *Discovery* days helped tipped the balance to ponies. Armitage had used ponies on the northern polar expedition of British explorer Frederick Jackson in the mid-1890s and was particularly enthusiastic about them. Indeed, Jackson had proclaimed the ponies 'an unqualified success'. However, that wasn't strictly true.

Jackson claimed that the ponies could carry extra loads, but much of the load was taken up by the feed, of which the ponies required large quantities. Another difficulty is that, because the weight-bearing hoof of a pony is four to five times greater than that of a dog's paw, ponies were far more likely to crash through the ice. Three of Jackson's four ponies died in this fashion. All of his dogs survived.

Royds, another acquaintance from the *Discovery*, had also recorded in his diary that he believed a mixture of ponies and dogs would have been preferable:

> Ponies can carry, or I should say drag, 1,800 lb a piece, whilst they eat, or require food, at the weight of 10 lb a day. A dog can carry 100 lb and requires 1½ lb per day, and a man can drag 200 lb and wants 2 lb a day food. So there is not much doubt that a combination between ponies and dogs with a small party of three men is the very best method of travelling over this snowy surface.

In an era when many of these decisions were still trial and error, Shackleton cannot be blamed for hedging his bets on ponies, especially after the disaster with the dogs on the *Discovery* expedition. That said, if the dogs had been professionally trained, they would have no doubt been far better performers. At least in this respect, Shackleton hired Bertram Armytage, a thirty-eight-year-old Cambridge graduate, to train the fifteen ageing Manchurian ponies he purchased from China. Armytage was, however, a curious appointment, as this was not something in which he had any experience. Just in case the ponies proved to be a failure, Shackleton also ordered nine Siberian huskies as some kind of insurance.

Ponies might have been a practical transport choice for the ice, but Shackleton was also aware of a potentially exciting alternative. Beardmore was manufacturing Arrol-Johnston motor cars at his Paisley works, and he realized that it could be excellent publicity for his company if Shackleton should drive all the way to the pole in one. It would, of course, also be the first such car ever used in Antarctica, an accolade in itself.

On the face of it, this seemed like a good idea. The car's

12–15 horsepower engine enabled a speed of 16 mph and its twin petrol tanks held enough fuel for 300 miles. Better yet, hot air from the exhaust would be diverted to warm the driver's feet. It promised a fast yet comfortable ride across the glacier, and towards history.

However, even the notoriously upbeat Shackleton was not taken in by this. He was aware that such a heavy vehicle, on tyres no less, would be useless on the ice, and no one knew how a car engine would function in extreme cold. It was for this reason that he did not pin his hopes on the car, and took ponies, saying to Skelton: 'I am not depending on the car as the main mode of traction.' But taking it along for the ride would certainly keep Beardmore happy and would also gather valuable column inches in the press, as seen in this article in *The Times*:

> The motor car which is to be used by Lieutenant Shackleton's Antarctic expedition is now on view at the Arrol-Johnston London depot before being shipped to Christchurch, New Zealand. The steel work in the frame of the chassis has been specially treated to make it resist the influence of low temperatures. The front pair of wheels are shod with wood, and provision has been made for the attachment of a sleigh, while the back wheels, similarly shod, have steel projections fitted in which holes have been drilled to receive spikes in order to obtain increased adhesive power. The engine is a 12 to 15 h.p. engine, air cooled, capable of giving about 16 miles per hour, and two systems of ignition are fitted. The exhaust is to be utilised for warming purposes, and is also connected with a snow melter, which will provide water. The two petrol tanks, one fed by gravity and the other by pressure, hold sufficient fuel for 300 miles.

One newspaper printed a cartoon of Jack Frost sitting on top of the pole, looking out over the ice, with the caption: 'Well, I've beaten dogs and ships, and balloons, and now they think they'll master me with petrol: Humph! We shall see!' Ramping up the hype, *Autocar* magazine quoted Shackleton claiming that, with the car, he would be able to travel 150 miles in twenty-four hours, and that 'there would be a fair chance of sprinting to the Pole'. Such a suggestion was of course hogwash, but the press and the public ate it up, and it won nervous glances from the Royal Geographical Society and Scott, who no doubt hated the idea of the motor car being used for such a noble endeavour. Shackleton was happy to play along to keep Beardmore happy, but the car would take up a chunk of the already limited space on the *Nimrod.*

While Shackleton was trying to cover all of his bases and thinking of experimenting with new methods if necessary, there was one area of preparation he shied away from: taking a medical. A concerned Emily eventually made him have a cardiology check-up, and while he said he had passed it, Emily was unconvinced. 'I think he examined the specialist instead of the specialist examining him! [in the words of a doctor who knew Shackleton well],' she said. 'I know the fellow never got a chance of listening to his heart . . . He may have been afraid that he wouldn't be fit to go.' As this was Shackleton's expedition, the only person who could stop him from going was himself, and there was no chance of that. He just had to pray that ill health wouldn't once more rear its head once he was on the ice.

Against all odds, by the end of July 1907, the expedition was ready to depart. It had taken Shackleton just a matter of months in which to raise all of the money, buy and refit a ship, recruit his team and obtain all the necessary equipment. This process took Sir Vivian Fuchs, whose two teams between them crossed Antarctica in the fifties, five years, and my own single team, who did the crossing in 1979/80 as part of our circumpolar journey, a total of seven years. But in racing against Scott, Shackleton felt he had no choice but to depart as quickly as he could. In the circumstances, while he was heavily debt laden, he did as good a job as anyone could have hoped to achieve.

Something that certainly would have saved him time was that in those days there were no Antarctic governing bodies, such as the Foreign Office Polar Desk and the US National Science Foundation to block non-governmental bodies wanting to operate down south. This saved Shackleton literally years of political wrangling. However, he did log his plans with the Admiralty, as well as with a special committee of Royal Geographical Society grandees.

The sight of the newly refitted *Nimrod* leaving East India Docks on 30 July 1907 must have stuck in the throats of Scott and the RGS. They had attempted all that they could to hinder Shackleton, but he had persevered and was once more heading to the Antarctic to the ripple of Union Jack flags, the blare of horns and thousands of cheers. But before he set off for the ice, he first had to make a couple of stops.

The first was at Eastbourne, so that the ever-loyal Elizabeth Dawson-Lambton could see for herself what her money had gone towards. While other, far wealthier patrons, had only gone as far as to guarantee loans, she had once more put her faith in Shackleton and got the ball rolling with hard cash. Without her initial confidence, the expedition might never have materialized.

The second stop was a few days later, at the Isle of Wight. The sea was crammed with over 200 ships, but the King and

Queen had eyes only for the small *Nimrod* as she sailed into port. Anchoring beside HMS *Dreadnought*, the mightiest warship in the world, Shackleton beamed with pride as King Edward presented him with the Royal Victorian Order. He had presented the same award to Scott six years earlier and, for Shackleton, this was a real sign that he had made it into the same upper echelons. He was no longer along for the ride. He was now the leader of a historic expedition, with the hopes of King and country resting on his shoulders. It was all he had ever dreamed of.

Then came a bonus Shackleton did not expect. So taken was Queen Alexandra with the expedition and the engaging Shackleton that, as a sign of her faith in him, she handed him a Union Jack flag to plant at the South Pole. Prickles of sheer pleasure crept over Shackleton as the graceful Queen announced: 'May this Union Jack which I entrust to your keeping lead you safely to the South Pole.' To add to Shackleton's delight, she had also attached a note to the flag which read, 'This is the first time it has been done.' Pictures of the *Nimrod* would subsequently appear in her new photo book, which aimed to raise money for charity, and it would feature on the Christmas cards she would send to friends later that year.

While the *Nimrod* set sail for New Zealand, Shackleton stayed on shore for a further two months to tie up loose ends. Eventually saying his goodbyes to Emily and the children, his mind was, however, firmly on the travails of the expedition preparations. Emily tried to 'make it as easy for him as possible', but Shackleton would come to regret his lack of attention to his wife pre-departure. When he did eventually set sail on a passenger ship, as soon as they had departed he wrote a litany of lovelorn letters saying things such as, 'It gives me a lump in my throat when I think of my family and I see the mothers on board ship with their children and I am going so far away from all I love,' 'I wish I was back in windy Edinburgh with you,' and 'I promise you again darling that never will I go away on this sort of thing again.' This was a familiar pattern. On shore, Shackleton showed little interest in his family, putting all his mind on expeditions, but once away from them, he missed them dearly. Either way, he was never fully content, always chasing something just beyond his reach.

One reason for Shackleton not being able to focus was the never-ending need for more money. With unpaid creditors still chasing him, including Beardmore, who wanted repayment of the £1,000 loan, Shackleton tried to parlay the King and Queen's public approval by trying to prise more money from prospective backers at Cowes Week.

Having managed to acquire some more funds, it was still nowhere near enough to be able to repay everyone. As such, Shackleton made the painful decision to leave without repaying Beardmore, something he knew would leave a bitter taste in the old man's mouth. He wrote to Elspeth in an attempt to explain: 'You cannot think much of me whilst I have not paid back to Will all the money he so generously lent me. But things have been very bad for me.' He subsequently tasked Emily's brother, Herbert Dorman, a solicitor, with handling all these outstanding financial matters while he was away, a thankless task indeed.

Still, he optimistically counted on being able to raise more money once he landed in Australia and New Zealand, which he could then send back to cover his debts. If all else failed, then he was hopeful that future book sales and lecture fees would put him back in the black. As he told Emily:

> I have already made arrangements with Heineman to publish the book on my return and it means £10,000 if we are

> successful. And that is quite apart from all newspaper news which we hope to fix up. It will leave me all the lectures etc, free and the book can pay off guarantees if the people really want them but I am of the opinion they will not ask for them if we are successful. I think it will be worth £30,000 in lectures alone.

All this would, however, depend on the success of the expedition, and with it being prepared in such a hurry, and Shackleton having fallen sick the last time around, that was far from certain.

17

A contemporary article in the *Field* magazine recently reported that, 'The primary qualification of the leader of a modern polar expedition is that he should be a good beggar.' Nothing has changed in that respect, but if Shackleton was to keep things afloat then he had to turn begging into an art form.

Even now that he was on his way to catch up with the *Nimrod*, and, as some biographers have speculated, possibly enjoying an affair en route, Shackleton was unable to keep his mind off the mounting financial pressure. He knew that as soon as he landed in Australia he would once more have to urgently raise funds if he were to keep the wolves from the door. If he failed to do so, there was still a very real possibility that the expedition would have to be scrapped altogether. This would be the ultimate humiliation. No matter what, he had to get more money.

Sure enough, as soon as he arrived, he found the situation more desperate than ever. Loans he was expecting to have been paid in the interim had failed to arrive, leaving him with insufficient funds to pay salaries to all of his men. Neither was there any money for the vast amount of additional food and gear that the *Nimrod* would need to take on board in New Zealand, never mind the various repairs that she would have to undergo.

Australia was, however, an untapped resource. With plenty of interest in him, and his expedition, Shackleton wasted no time in getting to work. Embarking on a successful lecture tour, Shackleton might have raised sufficient funds from the receipts alone. However, as he so often would throughout his career, he instead chose to donate the proceeds to local charities, earning plenty of goodwill along the way, but not solving his immediate problem.

However, while Shackleton may have had a chequered business history, he at least showed some sense in using the lecture tour as an investment opportunity. Lamenting the lack of support back home, he told his audiences, 'Indifference and the struggle to try and do something of which they did not approve has been our experience in England.' Swiftly pivoting, he then encouraged the Australians to prove that they were different to their miserly colonial masters: 'You have the enthusiasm for the glory of the Empire which keeps you from the condition of the people in the old lands, who are smoothing down into a state from which I hope the expedition will shake them.' In full flow, Shackleton was quite the showman, and in Sydney his salesmanship soon pricked the ears of a highly influential forty-nine-year-old Welshman.

William Edgeworth David was a renowned geologist who had discovered the Hunter Valley coalfields in New South Wales. Known far and wide, his opinion held sway in the upper echelons of Australian society as well as in government. David felt that Shackleton's expedition might potentially discover vast mineral deposits in Antarctica and therefore wanted to help. With this in mind, and knowing of Shackleton's financial troubles, he arranged for the Australian government to provide a grant of £5,000 (around £300,000 today) towards the scientific arm of the expedition. This was music to Shackleton's ears. In one swoop, he now had enough money to ensure that, at the very least, the *Nimrod* could set sail, even if creditors back home would remain unpaid.

Always with an eye towards further glory, and riches, Shackleton realized that if his expedition were to discover vast mineral deposits, then this would truly be his making. He and his crew might not be particular experts in this field, but who would be better than David to lead the way? While Shackleton could offer David only the meagre sum of £600 per year to join the crew, the famed Welsh geologist jumped at the opportunity. As Shackleton wrote to Emily, 'I think it is cheap to get such a man.' There was plenty of truth to this, in more ways than one.

David was soon joined by Douglas Mawson, a former pupil of David's and now a lecturer at Adelaide University. Their presence not only gave the expedition a raised profile in scientific circles, and expertise in discovering minerals, but just as importantly it meant that Shackleton could now focus solely on reaching the pole, leaving the scientists to do their work.

Armed with a formidable scientific team and sufficient money raised, Shackleton headed to New Zealand, to meet up with the *Nimrod* and her crew, all the while still looking to raise further funds.

In the meantime, some of Shackleton's associates had already been hard at work in Lyttelton. Joseph Kinsey, Shackleton's agent in New Zealand, had acquired the free use of the harbour for the *Nimrod*, while Leonard Tripp, a lawyer whom Shackleton had met while on the *Discovery*, had flung open the doors to New Zealand society. For Shackleton, this was an opportunity he did not want to waste.

Although Shackleton proved extremely popular in such circles, with the media also unable to resist the charm of the garrulous adventurer, he was left frustrated when it came to prising money from their pockets. Despite all of the warm words, New Zealand society was not as forthcoming with the cash as Shackleton had hoped. Once more, he was coming up short. But he still had one last, crucial card to play.

Upon being presented to New Zealand's Prime Minister, Sir Joseph Ward, Shackleton shamelessly pressed his case for the government to 'give a lead' in making a contribution. While the startled Ward considered this forthright request, Shackleton went to work on the beguiled press pack. Weaving a tale of woe, he told them that since being in the country he had 'interviewed fifty millionaires and not one would give me a penny'. Now he had been forced to ask the Prime Minister for a helping hand. The media reacted just as Shackleton had hoped. Fanning the flames on his behalf, one headline proclaimed, 'Will New Zealand Help?' In effect, Shackleton was strong-arming the government to cough up or be seen to be miserly. Not wanting to be seen to be on the wrong side of a man who was fast becoming a national treasure, the government eventually handed over another £1,000 to Shackleton's kitty.

The money raised did not scratch the surface in comparison to the amount of debt that needed to be settled back home, but it did at least allow Shackleton to fully stock the ship. If anything, they now had too much. Packed full with close to forty men, dogs, ponies and tons of supplies, the *Nimrod*'s Plimsoll line was nearly 2 feet under water. The ship was in fact so overloaded that there was not enough room on board to store sufficient quantities of coal for the journey. Despite all the effort Shackleton had gone to in purchasing the ponies, he now had no choice but to leave five behind in Lyttelton to make space. Even this sacrifice did not leave enough room to carry sufficient coal to power the *Nimrod* to the Antarctic and back to New Zealand.

To put it mildly, Shackleton was now in a bit of a pickle. However, he swiftly showed his talent for inventive thought and conjured up the idea of a larger steamer towing the *Nimrod* to the edge of the ice, therefore cutting down on the need to store so much coal. But this could prove to be expensive, and Shackleton was all out of money. Thankfully, Sir James Mills, the chairman of the Union Steamship Company of Dunedin, came to the rescue. Offering Shackleton the use of the 1,000-ton tramp steamer *Koonya*, he was also prepared to split the cost between himself and the New Zealand government.

Still, the search for funds was desperate. So much so that just hours before the *Nimrod* set sail Shackleton agreed to allow a local farmer, George Buckley, to join the team for the sum of £500. With time fast running out before departure, Buckley didn't even have time to pack the appropriate gear and so set off for the Antarctic with just a light-weight summer suit. It was yet another sign of the haphazard nature of the expedition.

Finally, on New Year's Day, 1908, the *Nimrod* left Lyttelton harbour, to the cheers of a 50,000-strong crowd and a brass band playing 'Auld Lang Syne'. Waving goodbye and lapping up the well wishes, Shackleton might have felt that the hard work had now been done. He could now concentrate on reaching the pole itself, which would surely be a breeze in comparison to all he had achieved in the previous months. Indeed, just before departing he wrote to Emily, 'In a few hours we will be off . . . and there will be some rest and peace.' Any yet, just hours after leaving port, the *Nimrod* was already in serious trouble.

18

Upon leaving port, the *Nimrod* was soon attached, via 9 tons of steel cable to the *Koonya*, which would lead her all the way to the Antarctic. However, the cable added even more weight to the already overladen boat, pushing its nose down further into the water. The Plimsoll line now disappeared entirely, leaving just 3 feet of freeboard remaining. Should there be any sort of storm, then the boat would be in severe danger of being overwhelmed. Shackleton was well aware of the risk, but at this stage had no choice but to pray the weather would be merciful. Yet just hours after their departure, they faced a hurricane force storm.

Waves over 100 feet high, roared on by ferocious winds, crashed into the ship. Rocking wildly on the towering waves, the *Nimrod* was forced at a 50-degree angle and in danger of capsizing. 'I have never seen such large seas in the whole of my seagoing career,' the *Koonya*'s second officer, Harbord, recalled. 'Water is coming on board by the ton.' It seemed only a matter of time before the two ships would be separated, with the chain being yanked and pulled in all directions.

Tons of ice-cold water swamped the *Nimrod* and flooded the decks, leading to real fears that the ponies would drown. The scientists were given two-hour watches to care for them, but there was not much they could do. Holding on for dear life themselves, they could do little but watch as the shrieking animals were tossed back and forth, smashing into the walls and collapsing to the floor. One of the ponies was so badly injured in the chaos it was unable to get to its feet. Now that it was of little use, Shackleton ordered it to be shot. Priestley wrote, 'If the ponies go, my private opinion is that our chances of getting to the Pole at all will go with them.'

The ponies might have been bruised and battered by the elements, while one of the dogs drowned, but the men on board didn't fare much better. Crammed into a tiny 15-by-8-foot compartment, they were thrown about, as if in a tumble dryer, with floors and walls sprayed with vomit, earning the compartment the nickname Oyster Alley. The dampness and the smell only added to the crew's already queasy stomachs. Marshall found the air to be 'foul', and wrote, 'one wakes up gasping for breath'. Douglas Mawson, as mentally and physically tough as any of the well-known characters of the 'Heroic Age', described their cabin as 'an awful hole'; 'we are wet all day by waves and we sleep in wet clothes between wet blankets at night'.

Priestley subsequently recalled the horrific conditions in detail in his diary:

> The so-called *Scientists' Quarters* is a place that under ordinary circumstances I wouldn't put ten dogs in, much less fifteen of our shore-party. It can be compared with no place on earth and is more like my idea of Hell than anything I have ever imagined before. To begin with it is big enough for only four men to live in under circumstances of relative comfort. There are no portholes that will open, the ventilation is pre-historic, and on two successive nights we have had to have all doors shut owing to seas being continually taken fair over the weather side of the stern. Every blanket in the place is wet through with salt water and the smell is almost insufferable. The floor is two or three feet deep in meteorological and other instrument cases, and in personal baggage.

Most of the men were too sick to eat or hold anything down but hot food was soon out of the question anyway. After a particularly heavy wave crashed into the ship, it swamped the galley and overwhelmed the stoves. As the storm worsened, members of the crew pointed fingers of blame at Frank Wild, who had killed an albatross during the voyage. Superstition dictated that he had brought these violent conditions upon them.

To much relief, ten days after the storm began, the weather gradually began to settle, with the first pontoons of pack ice now on the horizon. After towing the *Nimrod* over 1,500 miles, and being the first steel vessel ever to cross the Antarctic Circle, the *Koonya* could at last now relieve herself of her burden and make her way back to New Zealand. Shackleton would not forget the efforts of her captain, Frederick Pryce Evans. In as testing conditions as one could imagine, he had performed admirably. But now, at last, five years after departing the *Discovery*, Shackleton was back among the ice floes, and his adventures could truly begin.

Weaving and powering her way through the pack, the *Nimrod* was soon surrounded by icebergs on all sides. Many of the crew, who had never before set foot in the Antarctic, rushed up on deck and stared in wonder. Shackleton had of course seen it all before, but he was still invigorated by the 'indescribable freshness' of the landscape. In an instant, the stress of the last few months was wiped away. He was back, and doing what he loved best.

Although the weather had been fierce up until this point, the *Nimrod* at last struck upon some much-needed luck. Not only were the waters calm and the winds light, but Shackleton soon saw that rather than miles of pack ice to navigate there was now open water. Thanks to this sheer good fortune, by 16 January, the *Nimrod* had not only broken through the pack in record time but had also saved on coal. Yet now came the crux of something that had caused so much rancour in the previous months.

Shackleton had promised Scott that he would stay clear of McMurdo Sound and instead overwinter at King Edward VII Land. However, he now sought to land somewhere different altogether. He recognized that Balloon Inlet, where he and Scott had once risen to the skies in the hot-air balloon, was 100 miles closer to the pole than King Edward VII Land was, and as he told Emily, it would offer a 'straight road to the south and no crevasses'. With this in mind, Shackleton now changed course.

While the crew marvelled at the Great Ice Barrier and at the sight of hundreds of whales surrounding the ship, Balloon Inlet was nowhere to be found. It soon transpired that since the visit of the *Discovery*, the ice on the Barrier had changed shape, and Balloon Inlet no longer existed. Having made up so much time traversing the pack ice, they had now lost their advantage, and more. It was 24 January and the *Nimrod* was due to leave the continent on 1 March to avoid being trapped in the ice. This might seem plenty of time, but it would take weeks to land, unload supplies and assemble the huts. Time was already running short to do all that, but when they then attempted to reach King Edward VII Land it looked like ice barred the way. This left Shackleton in a quandary.

With all routes to his destination appearing to be blocked, an increasingly concerned Shackleton wrote, 'even if we eventually arrived at King E VII Land I might not be able to find a safe place to discharge and would probably have to abandon it in view of the enormous masses of land ice and hummocked pack that was breaking away which would make the ship's position untenable'. Nevertheless, Shackleton still vowed to keep his word and find a base other than McMurdo Sound, 'so that I could carry out my personal promise to Scott'.

However, Captain England was not only growing increasingly cautious about the conditions, he was also concerned about the shortage of coal; there was a chance they might not have enough to return to New Zealand. Recognizing the urgency, Shackleton told the crew to burn all available woodwork for fuel. The deckhouse, the mizzen mast and the topmast were all sacrificed, in the hope of buying the ship more time to find a place to land. Despite this, Shackleton soon realized that none of this would solve the immediate problem: beating the ice.

There was now only one option: McMurdo Sound. Shackleton knew full well what this would mean, and how he would be portrayed back home. He wrote to Emily, 'I have been through a sort of Hell . . . My conscience is clear but my heart is sore.' Yet he had little choice. He had tried two alternative landing spots and had been unable to reach either. If he were to keep trying, it would no doubt end in disaster. He might have had to break his word to Scott, but for the good of the expedition, and the safety of the crew, there was no alternative. While he told Emily that she could send parts of this explanatory letter 'to anyone you like in the family, Herbert, etc', she was not to send it, or discuss it, 'with outside people and to no enemies. Not that I think there will be really any except the Scott faction.'

In the same circumstances, I would definitely have done as Shackleton did. With all known routes blocked off, ice encroaching and fuel running out, he had no choice. If he had kept his word, he would certainly have not only endangered himself but also the crew. If all he would suffer was wounded pride at breaking a promise, then that was a very small price to pay.

However, not all of the crew agreed with his decision. While Harbord claimed that everyone was 'sick at heart and utterly disappointed', Marshall was outraged. 'Shackleton . . . hasn't got the guts of a louse,' he wrote, 'in spite of what he may say to the world on his return. He has made no attempt to reach K.E. Land . . . Got very angry when I told him I was sorry he had not made an attempt on K.E. Land. Tried to make me believe that he had done as much as any human being could.' In 1952, Marshall went even further, saying unequivocally, 'I have always been quite convinced that Shackleton never intended to land anywhere but at Scott's base.'

This is quite clearly untrue. Records show that Shackleton had tried hard to reach King Edward VII Land and that he was not helped by the ice or Captain England's caution. Recent research by Robert Headland, archivist of the Scott Polar Research Institute, supports the belief that Shackleton had always intended to land at King Edward VII Land and had to change course only because of circumstance. In addition, before setting sail for the Antarctic, Shackleton had persuaded the Prime Minister of New Zealand to appoint him as a postmaster. This would allow him to open the first post office in the Antarctic, from which his men could send letters home, via the returning *Nimrod*. For this, Shackleton had had 240,000 sheets of 1d stamps prepared, emblazoned with the name 'King Edward VII Land'. It is very unlikely that Shackleton would have gone to such trouble if he had never intended to land there.

Thankfully, most of the crew, including Captain England, understood and approved of Shackleton's decision. The very fact that he could genuinely claim to be as 'expert' at polar travel as anyone alive certainly gave him a great advantage in this respect. If he said the ship had to do something, then he was speaking from more experience in these waters than anyone else on board.

Nowadays, perhaps it is difficult to understand why this was such a difficult decision, but at that time, over a century ago, a person's promise was not lightly broken. And as far as Antarctica was concerned, a peculiar sort of ethics had grown up regarding rights and where people could travel. Jean Charcot, a French polar explorer, later said as much, stating, 'There can be no doubt that the best way to the Pole is by way of the Great Ice Barrier, but this we regard as belonging to the English explorers, and I do not propose to trespass on other people's grounds.' Amundsen, Scott's later rival, clearly agreed with this, for he wrote to Nansen, 'It is my intention not to dog the Englishmen's footsteps. They have naturally the first right. We must make do with what they discard.'

Broken promise notwithstanding, landing at McMurdo Sound still proved a challenge. Hut Point, where the *Discovery* had overwintered, was also now blocked by ice. As the month of January closed, Shackleton prayed that a storm would break up the 16 miles of bay ice and provide a path forward, but this was by no means certain.

In the meantime, as they prayed for the ice to break, Shackleton tried to make good use of their time. Adams, Wild and Joyce were instructed to man-haul to Hut Point, and reported back that Scott's hut was still in good condition. Shackleton also unveiled the Arrol-Johnston car, which he hoped could transport the stores over the ice to Hut Point in double-quick time. Aeneas Mackintosh's diary describes the 'mechanic' Bernard Day's chaotic first outing with his cherished machine:

> Alas! it kept up the alleged peculiarities of its kind, went a few feet and stopped dead, pulsating violently, until Day,

> moved no doubt by a feeling of pity, soothed it by a series of hammerings and screwing. After a brief rest, the machinery was started again, and the after-wheels in duty bound turned violently round in the snow, burying themselves to such an extent that the car moved not an inch.

The party hadn't even landed yet and already Shackleton knew this expensive machine, which was taking up plenty of space, was next to useless.

Of more concern was that Captain England's repeated attempts to ram the *Nimrod* through the ice proved fruitless. Time was fast running out, and Shackleton was growing anxious. Matters weren't helped when a heavy steel lifting hook swung free across the deck and smashed into Second Mate Aeneas Mackintosh's eye. Writhing in agony, he was carried down to the captain's cabin, where the doctors, Marshall and Mackay, applied chloroform, via a towel over his nose and mouth, and with a crudely constructed 'tweezer' fashioned from rigging wire removed the crushed eyeball by the light of a single lantern.

The mood was growing dark. It was rapidly becoming clear that the imminent arrival of winter would ensure that there was no chance of wintering at, or even reasonably close to, Scott's old base at Hut Point. Another landing site had to be found, quickly.

Sailing in and out of the ice, desperately searching for any site that might be suitable, Shackleton finally struck gold on 3 February. Able to access the tip of a rocky feature known as Cape Royds, named after the first lieutenant of the *Discovery*, Shackleton told England to get as close as possible so that he could explore.

Taking a small sailboat to the shore, Shackleton found a flat rock surface on which to set up camp. The spot was also sheltered from the cold winds by a nearby hill and there was a lake which would provide a regular source of water. It was more than Shackleton could have hoped for. Wasting no time, he immediately instructed England to bring the *Nimrod* closer in so they could begin unloading the stores.

However, England became extremely cautious. Fearful of crashing against the rocks, or of being trapped in the ice, he constantly moved the ship to shore and back again, wasting precious fuel. This allowed for some unloading, with the stores quickly dumped on the pack ice and then taken to land by heavy sledges, but it was nowhere near quick enough. Shackleton, who up until now had displayed due patience, was well aware that they had to complete the unloading with greater speed and less caution. Watching in barely concealed fury as England pulled the ship away once again, Shackleton was seen to tear off his hat, fling it to the ground and stamp on it.

Shackleton's patience with England might have been wearing razor thin, but the landing party was incandescent with rage. Chief Engineer Harry Dunlop came to the conclusion 'that the strain was too much for him. His nerves were in a bad state.' Frank Wild was far more scathing, claiming England had 'lost his nerve' and was 'off his rocker'. Mackintosh, meanwhile, wrote that England's yo-yo antics 'disheartened the whole party'. Matters were made worse soon after, when a storm carried the *Nimrod* 50 miles out to sea, leaving the shore party stranded, short of supplies and fearing they had been abandoned.

Shackleton recognized that this was the first time England had captained a ship, let alone in such rough and unpredictable waters. He tried his best to tolerate the situation, but an incident on the bridge, when the sea was calm, pushed him too far. Ordering England to take the ship to shore to unload more supplies, to his fury, England refused. Asserting his authority, Shackleton placed his hand on the telegraph and signalled 'Full Steam Ahead'. In response England changed the signal to 'Full Speed Astern'. England might have been captain of the ship, but Shackleton was his employer. This was an awkward balance, and it became obvious that the dynamic between them was fraught.

Retiring to their cabin so that they could talk in private, Shackleton asked England to resign on grounds of ill health, hoping that this would at least allow England to save some face. But England refused to step down. Shackleton now had to consider relieving him of command altogether, but he had to tread carefully. He didn't want to alienate England. Not only could he sour the group with his rants, but Shackleton still needed him to sail the *Nimrod* back to New Zealand. Moreover, legally, England could be removed from his position only when the ship was in port. Weighing all of this up, Shackleton decided that England would stay in charge, for now, but if he was to ignore his orders again, then Dunlop was instructed to put him 'in irons if necessary'.

After almost three weeks, and without further incident, all the ship's stores, equipment and animals were finally unloaded by 22 February. Sadly, the relationship between England and Shackleton was beyond repair.

When the *Nimrod* and Captain England left to return to New Zealand, Shackleton and his winter party barely said goodbye to them. Not only were feelings still running high after the pantomime at sea, but England had also insisted on a coal supply for the ship's return that was nearly 40 per cent over what was needed. This meant the shore party would have to skimp on coal during the harsh winter ahead. Shackleton had, however, ensured that this would be the last they would see of England.

England did not know it but, among the letters to loved ones he was taking with him, Shackleton had also asked him to deliver a letter to his agent, Joseph Kinsey. In the letter, Shackleton told Kinsey to dismiss England as an invalid on full pay for a year. If his eye had recovered in time, then the returning Mackintosh was to be promoted in England's place, and if this was not to be, then Captain Evans of the *Koonya* was to step in and return to Cape Royds in 1909.

Unsurprisingly, upon arriving in New Zealand, England was enraged by the decision, particularly the allegations that he had lost his nerve. 'I am perfectly sound both in body and mind and my resignation has been forced upon me,' he protested to Kinsey. Nevertheless, he had no choice but to accept it. Knowing that it could prove to be fatal to his career, he chose not to discuss it with anyone, not even the legion of journalists who beat a path to his door. But others who had returned on the *Nimrod* soon spilt the beans.

As Shackleton prepared to settle down for the winter with his party of fourteen men, he had no idea that, once more, he was front-page news all over the world. Having heard snippets of the power struggle between England and Shackleton, the press went to town, embellishing and exaggerating as best they could, knowing that Shackleton always made good copy. Newspaper headlines subsequently screamed 'Crew's Story of a Struggle on the Bridge' and 'Explorers Fight'. Columns were also full of lurid details of bitter dissension between the leader and his ship's captain, with there even being a mention of 'one of the combatants being knocked down'.

Further ill feeling was directed towards Shackleton when it was revealed that the *Nimrod* had failed to carry out a series of scientific experiments upon which grants from the Australian and New Zealand governments had been predicated.

Once more, Shackleton's word came into question. The *Lyttelton Times* was particularly scathing:

> The Commonwealth of Australia voted a sum of £5,000 to the expedition and New Zealand gave a further £1,000 in cash and much in kind on the distinct understanding that *Nimrod*, during the time the expedition was in the south, would return to Lyttelton and undertake exhaustive magnetic surveys of Australasian waters. These surveys would have been of more practical value to the Empire than even a successful dash for the pole, but they have not even been begun, and, it is understood, will not be undertaken at all. In the meantime, *Nimrod* is lying idly at Lyttelton, presumably awaiting the time when she will have to return to the Antarctic for the expedition. In view of these facts the public certainly have a right to ask whether the expedition intends to calmly appropriate £6,000 and leave undone the greater part of the unwritten contract for which the money was donated.

Creditors in Britain read the reports with fury. It seemed that Shackleton was duping more and more people, leaving behind a trail of unpaid bills and broken promises while raking in more money. Beardmore had also learned that his unpaid £1,000 loan had not even gone towards the expedition but to help Frank Shackleton escape his debts. From this point on, his relationship with Shackleton was at an end. But it was perhaps the news that Shackleton had landed at Cape Royds, within McMurdo Sound, rather than at King Edward VII Land, that promised to do the most damage to his reputation.

Predictably, Scott was enraged, telling Keltie, 'I cannot have any further dealings with Shackleton', and calling him a 'professed liar'. It seemed that all the news of fights on board the ship, unpaid creditors, money taken under false pretences and now this broken promise had vindicated Scott in his claim that Shackleton would do untold damage to expeditions for years to come. In the eventuality that, despite this chaos, Shackleton might miraculously reach the pole, Scott covered his bases, telling Bernacchi, 'I shall find it impossible not to doubt any result he claims. I am sure he is prepared to lie rather than admit failure and I take it he will lie artistically.'

The Royal Geographical Society recognized it was in an awkward position. Scott was their chosen 'action man', but now that Shackleton had landed in the Antarctic there was still a small chance he could reach the pole. The powers that be therefore chose to try to keep both sides happy. On the one hand, Keltie agreed with Scott and rubbished Shackleton, while on the other he told Emily, of her husband's decision to land at McMurdo Sound, 'Under the circumstances as described by him it was difficult to see that he could have done otherwise than he did.' Markham, Scott's greatest supporter, did likewise, claiming there were 'two sides to the story'.

In contrast, Wilson was entirely unforgiving. In a letter to a friend he later said of Shackleton's broken promise to Scott:

> As for Shackleton I feel the less said the better . . . I am afraid that he has become a regular wrong'un . . . In fact I have broken with him completely and for good, having told him in a somewhat detailed letter exactly what I thought of him and his whole business. I consider he has dragged Polar Exploration generally in the mud of his own limited and rather low-down ambitions.

For now, Shackleton enjoyed the luxury of being out of contact with the outside world, but there was little doubt that if he returned without reaching the pole, the vultures would tear him limb from limb.

Despite the stress and strain of the previous months, in his last words to Emily before he disappeared for at least a year Shackleton showed his determination to continue, no matter the odds, quoting their favourite Browning poem 'Prospice':

> Fear death? – to feel the fog in my throat,
> The mist in my face,
> When the snows begin, and the blasts denote
> I am nearing the place,
> The power of the night, the press of the storm,
> The post of the foe;
> Where he stands, the Arch Fear in a visible form,
> Yet the strong man must go . . .

19

On the desolate shore of Cape Royds fierce blizzards whipped up the sea spray and showered the men with shards of frozen ice. To help shield the men from these unforgiving elements Shackleton decreed that camp had to be set up quickly. Using picks to chip sheets of ice off the boxes, the men hurriedly erected the prefabricated 33- by-19-foot hut, which was to be lit by lamps powered by acetylene gas. In the darkness of the Antarctic winter, this would serve as their home for the next few months, in total isolation from the outside world.

Inside, the fifteen men were split into groups and each assigned a 6-by-7-foot cubicle with curtains to provide a modicum of privacy, particularly if they wanted to 'sport the oak'. The two Brit-born Ozzie scientists, Professors David and Mawson, were together, as were old 'polar dogs' Joyce and Wild, scientists Priestley and Murray, doctors Adams and Marshall, appointed pony-men Armytage and Brocklehurst, artist George 'Putty' Marston and motor-man Day, and cook Roberts and biologist/surgeon Mackay. Shackleton, meanwhile, shared his cubicle with the 'public library'. To make their cubicle their own, some chose to paint their curtains with suitable names, such as Rogues' Retreat for Joyce and Wild, alongside a cartoon of two rough types swilling beer from mugs.

If you were to nose around the hut, squeezing past the blubber-smeared cubicle curtains, taking in the foul-smelling air, you would find that down the middle of the central corridor was a table that could be raised on pulleys when not in use, to provide a little more space. The most vital item, the coal- (or seal blubber-) fired stove, which was both an oven and a central heating source, squatted at the end of the corridor, away from the entry porch and next to the cook's work table. A 'fridge' area was subsequently assigned along one side of the hut for seal and penguin corpses, which were regularly garnered from the nearby frozen lake, as if it were a local supermarket, with over a hundred killed on one particular day. The 'fridge' was, however, kept far away from the kennels on the other side, with the dogs ready to snatch any scraps they could.

With the camp set up, there was still much to do and, as always, Shackleton led from the front. Helping out in all areas, he could be found shovelling coal, exercising the ponies, or acting as the night watchman. This was typical of Shackleton, helping to ensure that there was no class divide and that everyone mucked in. He was also good company. Unlike Scott, Shackleton was never shy in public, nor when parties became rowdy in the mess. As the hut shuddered at the might of 80 mile per hour katabatic winds, with fine powder snow squeezing inside the tiniest holes or cracks in the walls, the men would gather along the narrow gas-lit table, their stark shadows mingling, ghost-like, as Shackleton held them enraptured. 'He was a sociable man and liked company,' Priestley remembered, 'and was always the life and soul of any group in which he happened to be.'

However, it was perhaps the printing press that gave Shackleton, and the men, most joy. The *South Polar Times*, as mentioned, had been a huge success on the *Discovery*, but this time around Shackleton wanted to publish a book for the party to read, with everyone contributing. It was to be titled *Aurora Australis*, with Marston producing the artwork and Shackleton again using the pseudonym Nemo. Before setting off, Wild and Joyce had undertaken courses in typesetting, compositing and printing, so they were in charge of producing a hundred copies of the 120-page book, while Day worked the Venesta cases into covers. When complete, the book proved very popular and certainly helped keep the party entertained.

But it wasn't all fun and games during the dark winter months. When Marston sought to entertain everyone by dressing up as a woman and proceeded to make comic advances around the room, Mackay erupted and the men had to be separated. Worse was to follow, as was detailed in Marshall's diary, when Roberts, the cook, and Mackay, fell out over Roberts putting his feet on Mackay's sea chest. Mackay tried to throttle him and, while the burly Marston managed to subdue him, Shackleton knew he had no choice but to lay down the law. To let such a matter go could seriously undermine the group and put the expedition in jeopardy. In a rare flash of temper, he threatened to shoot Mackay if he were to repeat such behaviour.

Shackleton was clearly a rare breed who could lead men from different backgrounds, whether hardened seamen or scientists, and earn their respect, by force of personality or by strength. Buckley recalled that he had a 'magnetic influence and soon he was known to all as "the Boss"'. But the Boss now had something of a headache.

Thanks to the super-cautious England, the delay in unloading meant Shackleton was unable to start laying supply depots on the Barrier for their planned spring march to the pole. Ice, which had primarily blocked the route, had now floated out to sea, and until McMurdo Sound iced over again there was no way to reach that point.

David, however, had a suggestion that caught Shackleton's attention. The nearby 13,280-foot-high Erebus volcano had never been climbed. Rather than be confined to the hut all winter, a small group could scale it and immediately put the expedition in the record books. It was also a chance for David, Mawson and Mackay to get some experience, as the trio were expected to launch a later expedition to locate the South Magnetic Pole.

After much discussion, it was agreed that David, Mawson and Mackay would spearhead the attack, while Adams, Marshall and Brocklehurst, the young baronet, would offer support from behind. However, none of the men had any prior climbing experience, nor any real climbing equipment. All they had to scale the volcano were crude home-made crampons (bits of leather studded with nails attached to their smooth-soled finnesko boots) and an ice axe. Still, there were no real concerns, as the ascent slopes seemed to offer a reasonably gentle climb.

Towing behind them a 600-lb sledge loaded with provisions for ten days and two three-man tents, the men cheerfully set off, hoping to make their mark. Things were, however, made more difficult soon after, when the sledge broke and each man had to share the load in makeshift rucksacks on their backs. As the tent poles were too cumbersome to carry, they had to be discarded, which would make it impossible to erect the tents for shelter. Rather than turn back, and showing incredible naivety, they vowed to just use the tent cloths to cover them while they were in their sleeping bags. The omens were already not looking good.

Making slow progress up the volcano, the team began to realize that they had bitten off more than they could chew. The crampons were ineffective, leading them to constantly slip and fall, windmilling their arms to keep balance, and their hands were numb, unable to grip. 'We were struggling for some time,' David wrote, 'mostly on our hands and knees . . . no breath for talking.' Scaling ever higher, Mawson had a bad dose of altitude sickness while Brocklehurst suffered from frostbite on his feet. Such was his condition that the group had no choice but to leave him behind and carry on climbing, promising to pick him up on their return. It was quite a way to mark what was his twenty-first birthday.

Five days after setting out, the exhausted group reached the summit of the hissing crater spewing steam and sulphur and were now able to see across a vast stretch of the ice shelf. Despite all the travails, it had been worth it. They were in the record books and had also gathered some vital experience on the way.

After picking up Brocklehurst and some geological samples, the group decided that the quickest way to the bottom would be via glissading – a polite word for bum-sliding. Putting their axes between their legs to act as a break, they pushed off and careered down the slope, whooping and hollering all the way. Soon they were safely back at Cape Royds, accepting back slaps and congratulations. Exhilarated by his men's achievement, Shackleton cracked open the champagne on their behalf, claiming that it 'tasted like nectar'. At the very least, his expedition would have this first assent feat to show for it.

However, he didn't lose sight of the real reason for the expedition: to reach the pole. Not only would that cement his status as a hero and earn his fortune, it would also silence his many enemies. Yet that task was made far harder when four more of the ponies dropped dead. When Marshall completed a post-mortem, he found that their stomachs were full of volcanic sand. For a week or so after landing, while their stables were being made ready, the ponies had been tethered on a sandy patch not far from the hut. It now seemed that they had taken to eating the sand, due to hunger for salt. After originally setting out with fifteen ponies, only four remained. This was a disaster. Shackleton's plans dictated that he needed at least six ponies to go with the five men he had already chosen to join him on the march: Adams, Brocklehurst, Joyce, Marshall and Wild. He felt that four ponies might still work, but if any more were to die, he would have a real issue to contend with.

While the ponies were a concern, Shackleton's chosen team members were also falling like flies. The frostbite on Brocklehurst's big toe had turned gangrenous, and needed to be amputated, which put him out of contention for the journey south. This was not too much of a blow, as Marshall had already cast doubt on Brocklehurst's suitability when he noted he had 'an irregular heartbeat'. In his diary, he also offered his opinion that Brocklehurst 'seems to have no guts'. In any event, Shackleton duly gave up his private cubicle so that Brocklehurst could recuperate.

Soon after, Joyce was also ruled out, when Marshall found that his liver had been damaged by alcohol and that he had symptoms of heart disease. Shackleton knew only too well that a sick man on such a journey could put everyone's life at risk. Yet it was for this danger that his own participation came into question.

When Marshall performed a health check and listened to Shackleton's heart, he declared, 'Pulm[onary] systolic murmur still present.' Marshall was not surprised by this. While Shackleton's health issues on the *Discovery* were well known, Marshall had also suspected that Shackleton had avoided climbing Mount Erebus as he was feeling the effects of the cold and high altitude. As Marshall was to recall in 1956, 'he knew he could not have stood the altitude, likely to be over 12,000 ft and would have proved his incompetence before the southern journey'.

But there was no way in hell that Shackleton was going to rule himself out of the trip south. Marshall tried in vain to reason with him but by now he had come too far and risked too much. Yet if Shackleton was going to go south with such a health condition, then he knew he had to make doubly sure that all those who travelled with him were healthy. One man down was bad enough; two would be a total disaster.

Now, with Brocklehurst and Joyce ruled out, Shackleton spent the night deep in contemplation. After hours of wrestling with his options, he decided that the party that would travel south would be himself as boss, with Adams as sub-leader, Marshall as doctor and the ever-loyal Wild, owing, when all was said and done, to his personal standing with Shackleton. It was still a strong team, capable of reaching the pole, as long as no more ponies died and none of the men had health issues.

The health of the remaining men was, thankfully, not in question, but a personality clash came to the fore in the four months of polar darkness. Marshall, a highly educated and opinionated man, soon came to resent Shackleton being leader. He apparently believed that with his university degree and higher level of education, he deserved more respect from Shackleton, and no doubt the crew. Shackleton was, however, able to lead the men through personality alone; he didn't need to rely on having a level of education. His leadership was a natural gift, which few possess, certainly not the insular and moody Marshall.

Unable to overcome his jealousy, Marshall made veiled criticisms whenever the chance arose, while also ignoring Shackleton as best he could. In private, he spewed his bile into his diary, writing of Shackleton soon after the Erebus climb, for no known reason, 'He is incapable of a decent action or thought . . . Shall get my own back before I have finished.' Another subsequent, and similarly derogatory entry railed about Shackleton kicking the dogs and not holding regular church services, concluding with Marshall calling him a 'consummate liar and a practised hypocrite', and not having 'one iota of respect for him'.

In the darkness of winter, Shackleton kept his distance and hoped Marshall's mood would improve when the light returned. If it didn't, then he faced the difficult but necessary decision of leaving Marshall behind. But could he trust him not to pollute those who would remain with him?

When the light returned in August, the real preparation for the journey south began. Of most importance was laying the supply depots, which had been postponed due to the delays in finding a landing site. On 22 September, Shackleton, Adams, Marshall, Wild, Marston and Joyce set off, with their goal to reach Hut Point and establish their first depot at 79 degrees 39' S. It was a 320-mile round trip, a true test for all involved, particularly as Shackleton planned to watch Marshall closely.

For the trip, Shackleton decided against using the ponies. This might have made their task harder, but he was worried in case any of them suffered death or injury before the big push south. He did, however, give the Arrol-Johnston car one last try to see if it could be of any use in carrying them or the equipment. Alas, after stuttering and starting, it gave up the ghost after just 8 miles. Not only was the lack of tracks on the tyres fatal to its performance, Day also found that the extreme temperatures were choking the carburettor's inlet jet. Seventy-six years later, in similar conditions, our one-man snow scooters had the same problem with temperatures in the minus 50s: throttles jammed, carburettors blocked, fuel lines leaked and a gearbox gasket blew up.

There was now no doubt in Shackleton's mind. The car had been a gigantic waste of time. While Day later recorded a speed of up to 15 mph carrying a sledge load of 750 lbs, moving in eight hours what a six-man team would have taken at least two days to haul, it was all on hard ice. If only the car could cope with soft snow . . . If only pigs had wings . . . For now, the men would have to complete the rest of the journey man-hauling.

After tidying up Scott's old hut, they pressed on to the Barrier, where the temperature dropped to minus 59 degrees Fahrenheit and the wind picked up significantly. It was a true test for the group, especially when Joyce developed frostbite. The conditions became so bad that on some days they could not even leave their tent. Despite this, they still managed to achieve their goal, reaching 79 degrees 39' S, where they dumped oil and pony fodder, marking the area with a black flag.

On their return to Cape Royds, Wild wrote, 'We were a most ravenous and weary party on our arrival.' Meanwhile, David, Mawson and Mackay had already set off, hoping to locate the South Magnetic Pole. It was a round trip of 1,000 miles into the unknown, and Shackleton had told them that if they were to meet up with the returning *Nimrod* and return to New Zealand, they must get back to Cape Royds by 7 February 1909.

Yet to Shackleton all of this was merely just window dressing for the main purpose of the expedition. At the end of October, he and his team would finally set out to reach the pole, but with just weeks until their departure, the question of Marshall's temperament still remained. He had been an almost malign presence in the hut, but Shackleton found he had worked admirably on the ice and had no further issues with him. Indeed, Shackleton was reluctant to leave the ultra-tough Marshall behind, despite other doctors being available.

On his previous attempt to reach the pole, Shackleton had almost lost his life to scurvy, and one of the key reasons

Marshall had been chosen for the expedition was his study of the disease. Although the lack of vitamin C had still not been identified as the cause, Marshall's research had led him to believe that the only way to counter the disease was by eating plenty of fresh food, primarily seal and penguin meat, as well as tinned tomatoes, which he had acquired in New Zealand. The tomatoes, in particular, held good doses of vitamin C, which could keep scurvy at bay and even help cure it. Shackleton was therefore relying on Marshall's expertise to help the party to the pole and back, without scurvy rearing its head. Unless Marshall proved incapable of controlling his bitter moods, Shackleton remained determined that he would join them.

This goes somewhat against my own experiences. I learned over time that if I had any inkling before departing that someone might be a bad apple, it was better to get rid of them immediately. The ice is hard enough without someone souring the mood and constantly challenging you. However, while Shackleton put Marshall on probation, Marshall still wasn't convinced that Shackleton should go at all.

Shackleton might have shown no ill effects while on the depot-laying operation, but Marshall was sure that his heart murmur would become a problem when they were further out on the Barrier. The two soon butted heads over the issue, but Shackleton made it clear that he was the 'Boss' and if he was determined to go, then there would be no stopping him. The matter was only resolved when, shortly before the start of the journey south, Shackleton gave Marshall a formal letter, exonerating him from any blame or responsibility for any health problems which might arise on the march. It also declared that if Shackleton failed to return, his executors would defend Marshall's medical judgement. Shackleton wrote, 'I have accepted all your medical opinions and acted accordingly,' and signed the document. Yet the decision to let Shackleton continue haunted Marshall for decades afterwards. Writing half a century later, he concluded that Shackleton was 'Never fit to carry out any of his programmes.'

Before leaving England, Shackleton had made Emily a solemn promise: 'Of one thing I am certain, I shall not run any risk for the sake of trying to get the "Pole" in the face of hard odds.' But now, in a letter to be handed to her in case of his death, he wrote, 'My own darling Sweeteyes and Wife. Think kindly of me . . . Remember . . . your husband will have died in one of the few great things left to be done.' No matter what, Shackleton truly believed his destiny was to reach the pole, or he would die trying.

20

On 29 October 1908, Shackleton's moment of truth finally arrived. Marshall, Wild and Adams, as well as a handful of others from the camp, led by Joyce, who would join them for the first few days of the march, set off under a clear blue sky. It was, as Shackleton said, an 'auspicious beginning'. Bernard Day drove the motor car on the hard ice for the first few miles, to help carry supplies, while the ponies, Chinaman, Grisi, Quan and Socks, each towed a sledge, with none of the dogs making the journey. Scott would have approved.

Before departing, Shackleton told Murray, who was left in charge at Cape Royds, that if they had not returned by 1 March 1909, 'then something very serious must have happened'. In that eventuality, the *Nimrod* was therefore ordered to return to New Zealand, with a party of three left behind at Cape Royds for another year, on the off chance that Shackleton and his men would return.

Yet Shackleton carried no thoughts of failure in his mind. Ever since he had been told to return home by Scott almost six years earlier, he had done everything in his power to get back to the ice. In the interim, he had overcome scandals, embarrassments, back-stabbing, outright hostility and sabotage, and yet, somehow, here he was, determined to finally claim what he believed was his rightful destiny: a place in the history books. As he wrote himself, 'At last we are out on the long trail after 4 years thought and work. I pray we may be successful for my Heart has been so much in this.'

Shackleton knew better than anyone that the terrain ahead

was treacherous. Alongside Scott he had reached 82 degrees 15', well short of the pole, in a march that had almost killed him. Now he aimed to go much further. On his last attempt he had walked the first 300 miles of the journey, but the remaining 560 miles to the pole was uncharted. It was impossible to say what challenges lay ahead. The path to the pole could be blocked by mountains the size of Everest, it could be located in a network of deep canyons that stretched for miles or, like the North Pole, it could be on sea ice, cut off from land.

To achieve this potentially impossible mission, Shackleton had ninety-one days before food supplies ran out; Scott had only taken enough for seventy days. To complete the round trip of 1,720 miles, Shackleton and his team needed to travel at least 19 miles a day, on a daily ration of just 4,300 calories. On the face of it, this seemed a ludicrous goal. Not only was this twice as fast as they had previously managed on the *Discovery* trip, but in ninety-three days they had marched only 960 miles, well off the pace required to reach the pole. Despite such monumental odds against them, Shackleton remained optimistic that they could somehow achieve their mission. While he was confident that the remaining four ponies would be of great assistance, he also knew better than most that force of will and strength of character can see a person through the most testing conditions.

This optimism and determination clearly rubbed off on his men. Brocklehurst recalled that 'Shackleton was so enthusiastic and so confident in his own ability that he didn't leave very much for us to think other than success.' Shackleton was a man of many faults, but when it came to leading by example and making a team a greater sum than its parts, he had few equals. On the ice, these characteristics would be crucial.

Indeed, Shackleton quickly faced his first test when Socks went lame. Soon afterwards, Adams suddenly let out a cry. Quan, the largest of the ponies, had kicked him in his shin, cutting the flesh to the bone. With Socks lame, and Adams barely able to walk, Shackleton had no choice but to head back to Hut Point and postpone the march for a few days, hoping that Socks and Adams would recover. Frank Wild, loyal fan of Shackleton though he always was, recorded, 'The skipper is rather irritable and excited.' To Shackleton's great relief, by 3 November both were fit enough to continue. Yet as the party again set off on to the Barrier, further challenges quickly followed.

After much consideration, Shackleton had bet on ponies over dogs to help carry provisions and equipment to the pole. But as the animals stepped on to the soft snow, as had been feared, their weight saw them sink below their knees, dramatically slowing the march.

The ponies fared little better on the ice. Unbelievably, they weren't shod, which meant they constantly slipped and fell. Scott was later to note the excellent results of easily fitted snowshoes for ponies in soft snow conditions, but I can find no record of Shackleton's team even considering using them.

To help the distressed animals, Shackleton and his team cut steps into difficult stretches of ice, but this took up crucial time. Suddenly, achieving 19 miles a day didn't just seem optimistic, it was nigh on impossible.

The conditions were also difficult for the men, as they also sank into the soft snow. On foot, the men exerted pressure of 2.5 lbs per square inch, while on skis the pressure would have been as low as 0.5 lbs. However, as has been noted, skis back then were very heavy and would only be required in certain circumstances. It was therefore a calculated risk not to take them. If they were not required, the men would be lugging extra weight for thousands of miles. But if they were needed, they would be missed, as they most certainly were at this point, with the men already well behind schedule just days into the march. Shackleton soon realized he had to act, and fast.

At this speed, their daily rations were going to run out before they even got close to the pole. Shackleton therefore made the decision to cut them to last 110 days, which would allow the men to aim for 12.5 miles a day. He hoped this might be a more realistic goal, stating, 'If we have not done the job in that time it is God's will.' Hedging his bets, just in case of any disasters, when the supporting party turned back on 7 November, Shackleton told Joyce to lay an additional food depot at Minna Bluff. If things were truly desperate by then, this would act as insurance.

First and foremost in Shackleton's mind was making the required distance each day, but he was also still wary of Marshall; particularly that he would poison the minds of the others against him. To avoid any discontent, he decided that each week the men would switch tents so that anyone with a grievance would have little time to form a clique to challenge his authority. He also ensured that all duties were shared, such as cooking, and that he included the men in discussions about key decisions. He hoped that this would make them feel that they were all responsible for the success or failure of the mission and invoke team spirit. They were all in this together, come what may, responsible for any success, as well as any failure. If he tried to assert his authority in such treacherous conditions, he feared that the men might snap and blame him for all their ills.

Shackleton's hopes were, however, dealt another blow. While they had covered 54 miles in the first nine days, bettering the 37 miles they had covered in the same period with

Discovery, a fierce blizzard set in, confining them to their tents until the storm passed. With every mile and every day vital, this was an enormous blow. The party had no choice but to shelter in their tents and hope the storm passed.

The tents shook at the battering of the ferocious wind, while each man huddled inside, reading books that they had taken with them, in Shackleton's case, Shakespeare's *Much Ado about Nothing*. Being unable to march was one issue, but in the meantime the men and the ponies were also consuming valuable provisions over the two days they were pinned down. An unspoken fear now pervaded the tense atmosphere: the pole was already out of reach. While, in private, Shackleton wrote, 'It is a sore trial to one's hopes,' he knew that he could ill afford to have the men accept defeat so early on, even if everything seemed against them. He did all he could to remain optimistic, and show that he was relaxed, cracking jokes, telling stories and taking an interest in each man to keep their spirits high. This seemed to rub off on the men, who believed that if Shackleton thought their goal was still possible, then all was well.

By 9 November, the conditions had settled enough to allow the march south to resume. But the going was still tough, particularly when they encountered frozen waves of sastrugi, which were treacherous underfoot. Moving tentatively forward, Shackleton wrote, 'There was nothing for it but to trust in Providence.' And yet Providence seemed determined to desert them. As Chinaman stepped over a crevasse, the ground gave way leaving him dangling over an abyss, with him carrying all of the camp's cooking gear, half the oil supply and all of the biscuits. Racing to his side, the men desperately propped the pony up, all the while aware that the ice could give way under their feet at any moment. To lose Chinaman and the cooking supplies would not only make the march ahead impossible, it would also put into question any return to Cape Royds. Using all their might, and straining every muscle, they managed to haul the pony to safety.

The ponies were proving to be more trouble than they were worth. At camp each night the men realized that, unlike dogs, who retain their heat, the ponies dissipated it, so they were particularly susceptible to the cold conditions. To keep them warm, the men had to erect windbreaks made out of snow blocks, wipe off any snow from their coats and hooves and then strap on night blankets. In addition, their maize needed to be boiled into a hot mash. Dogs would have been far easier to care for. Soon the men came to hate the ponies, but they had no option but to continue with them.

Thankfully, conditions at last improved. With the ice becoming firmer underfoot, the men not only reached the daily target of 12.5 miles on some days, but on others surpassed it. All being well, they might yet just reach the pole. Even the irritable Marshall wrote, 'Everything more hopeful.'

Having reached the first depot to refuel on 15 November, the now well-fed party continued to claw back further valuable time in the week that followed, meeting and exceeding their daily target. On 16 November, they achieved a daily record of 17 miles, reaching the 80th parallel. 'I am beginning to think we shall get to the Pole alright,' Wild wrote, although he added pessimistically, 'but am doubtful about getting back again.'

The health of Chinaman didn't help matters. Badly ailing, with cut fetlocks, it was clear that he could no longer continue. Wild was called upon to perform a sacrificial duty that neither Shackleton, Adams or Marshall could bear: to shoot the pony in the head and end its misery. Marshall at least had no such reservations about carving up Chinaman's meat, with 50 lbs to be left at the next depot for their return march. The rest was heartily devoured that night, either fried or chewed raw. Wild said it tasted like good beef and he downed 'quite a respectable amount', while Marshall commented that the meat should help them 'lose the danger of scurvy'. In eating it, Adams had trouble with a tooth, which Marshall, with considerable difficulty, and no forceps, managed to extract.

Although Chinaman's meat would no doubt prove to be crucial on their return, the party was now a pony down, and its load had to be shared among the others. To lose a pony so far into the march wasn't disastrous at this stage, but it certainly would be if any more were lost in quick succession.

Remembering just how difficult it had been to locate each lone black depot flag during the *Discovery* expedition – which Shackleton had said was like 'picking up a buoy in the North Sea' – he now decided to make the depots far more conspicuous. Erecting a mound of snow with a black flag placed on top, the men built additional mounds of snow 6–7 feet high, which would stand out in the flat landscape and stand a better chance of not being wiped away by bad weather.

Now moving further east to avoid the broken and moving ice which had caused the *Discovery* expedition so much trouble years before, Mounts Longstaff and Markham came into view. Yet before them was something far more important. Scott's furthest south record of 82 degrees 17' was now tantalizingly within reach.

Just days later, they surpassed it, when on 26 November the men camped at 82 degrees 18.5' S. It had taken Scott, Shackleton and Wilson fifty-nine days to set their record. Shackleton and his men had broken it in just twenty-nine. This might not have been Shackleton's ultimate goal, but whatever was now to come, he could be satisfied that not only was he a record breaker, but he had proven himself to Scott and all those who had doubted him.

All that now lay before them was uncharted territory. 'It falls to the lot of few men to view land not previously seen by human eyes,' Shackleton wrote in wonder. 'You can't think what it's like to walk over places where no one has been before.' However, he was well aware that in the thrill of new discoveries, the march ahead might throw up many unexpected and unwelcome challenges.

For two weeks, the men had made miraculous progress, and all were still in good health, but then both Grisi and Quan became lame. There was no option but to shoot them, with Grisi's meat to be left behind at the next depot. Marshall blamed Shackleton's 'ignorance and incompetence' for the death of Quan, the old enmity between them simmering just below the surface.

The men now had to share the ponies' workload between them, adding another 600 lbs, with the pole still 400 miles away. With rations already cut, this only increased the men's already ravenous hunger. As energy reserves rapidly depleted, so did the distance they covered. While they were originally rationed to 4,300 calories a day, they were now consuming no more than 3,000. 'We are very hungry these days and we know that we are likely to be so for another three months,' Shackleton wrote, recognizing the scale of the task ahead. 'Shall be glad to get out,' Wild, now sharing Shackleton's tent, wrote in cipher.

If battling hunger wasn't enough, what came next was a significant blow. In order to reach the pole, it was vital that the surface ahead remained flat and solid. Yet before them now was a chain of mountains known today as the Transantarctic Mountain Range. This feature stretches some 2,200 miles long, is 180 miles wide and in places over 14,000 feet high, just the sort of uncharted territory Shackleton had feared.

Just getting to the foot of this mountain range was a challenge. The pressure of ice against rock had forced up huge ridges, making the terrain extremely dangerous. It was, according to Marshall, 'like the ocean swell breaking on an unknown shore'. There seemed to be no easy path through, and the men worried they had come as far as they could go. Marshall wrote, 'I was here asked to tell the Boss he "had done damned well and had done enough." I replied "there is nothing to stop anyone turning back."' Nevertheless, Shackleton refused to admit defeat. Instead, he decided to climb the nearest high ground to see if there might be an easier path ahead. If there was they would continue towards glory. If not, they would have no choice but to turn back. The men, who now all called Shackleton 'Boss', chose to put their trust in him.

Tying ropes to each other, the men tentatively navigated their way through 7 miles of crevasses. Danger was on all sides. Any slip could send a man tumbling hundreds of feet below. Unsurprisingly, progress was agonizingly slow, with the men taking six hours to cover 7 miles, before reaching the foot of a red mountain.

Compared to others in the range, the mountain was at least relatively small, being just 2,000 feet high. Within two hours, the men had neared the summit, praying that the view from the top might show a more welcoming path on the other side.

Shielding their eyes from the sun, they peered into the distance to see a giant glacier forging a southerly path that penetrated the mountain range, like some highway to heaven. 'There burst upon our vision an open road to the south,' Shackleton excitedly wrote. If they could just get past the crevasses, with Socks and their provisions, their path to the pole was clear. Even Marshall took a moment to recognize the magnitude of their discovery. 'Shall never forget the first sight of this promised land,' he wrote. 'The Almighty had indeed been good to us.' Shackleton had discovered one of the few negotiable routes through the mountain range and to the pole.

Every new stretch of uncharted territory also provided the opportunity to honour a prominent local feature with a name. Shackleton therefore opted to call the red mountain Mount Hope. While some might have believed that 'Hope' was meant to signify his feelings upon seeing the path to the pole, there appears to be another explanation.

Years before, Shackleton had met a Scotswoman named Hope Paterson. Again, there is no evidence that a romantic relationship occurred, but we certainly know from letters that the two remained in close contact. Upon Shackleton's return home from the expedition, he sent her a piece of rock, held in a silver case, which had on it inscribed the map coordinates for Mount Hope.

There was also some debate surrounding the name Shackleton gave to the glacier. Most assumed the name Beardmore Glacier was in honour of William Beardmore, in an attempt perhaps to pacify him after Shackleton had disappeared without settling his £1,000 debt. Notwithstanding this, Shackleton had previously told Elspeth Beardmore that he would name a glacier in her honour. It has therefore long been argued by some that perhaps it was actually named after Elspeth, not her husband.

At this moment in time, such names were trivial matters. Shackleton had already broken Scott's record, and now he had miraculously found one of the very few gateways available to the pole. He would not, as he had feared the previous day, be returning home with his tail between his legs. Suddenly, all of his dreams seemed within touching distance.

21

Having descended the mountain, they returned to camp, the sun shining brightly in the sky, as if heralding the group's momentous discovery. After weeks of hard toil in the bitter cold it seemed to mark a change in fortunes. But this was not to be.

Hauling a load of 600 lbs between three, while Socks, led by Wild, pulled another, was hard work, and the blazing sunshine didn't make it any easier. It was now midsummer and the sun shone even at midnight. Sweating profusely, the men had to strip down to their shirts and chewed chunks of frozen horse meat to help keep them cool.

The sun also played havoc with their eyes. As they traversed crevasse bridges and scaled Mount Hope, the sun's rays reflected off the ice and blinded them. Their snow goggles also fogged up with perspiration, while their slit eyes, puffed up with fluid, streamed with tears, freezing their face masks to their cheeks. Refusing to wear his goggles, so he could navigate a path ahead, Shackleton suffered more than most, with the sun's ultraviolet rays burning not only the corneas in his eyes but also any exposed skin, with his face becoming pocked with blisters and scabs, while his lips cracked and bled. To try to ward off the pain he pressed his eyeballs with his mitts, but this didn't help much, and each time he tried to open his scabbed lips to speak, he tore further strips away from the skin, making his lips bleed then freeze. Wild noted that snow-blindness was giving Shackleton a 'fearfully trying time', but admired the fact that 'he would not give in'.

On 4 December they finally set foot on the giant glacier, the 'Gateway' to the pole, but Shackleton was aware that the sun's heat might well cause many of the crevasse bridges to thaw and collapse beneath their feet. He was particularly wary when one fragile snow bridge cracked and creaked at each of their steps, seemingly threatening to collapse at any moment.

Shackleton also recognized the threat from above. Surrounding them were huge boulders that had fallen from the surrounding cliffs. He wrote, 'One feels that at any moment some great piece of rock may come hurtling down.' At night this was a particular concern. Unable to anchor their tents on the hard ice, they had to do so on a patch of soft snow instead, which lay under a granite pillar, with discarded rocks all around. 'Providence will look over us tonight,' Shackleton wrote, 'for we can do nothing more.'

They soon realized it was too dangerous to carry full loads across the fragile and treacherous ice. They therefore decided to leave Socks behind while they relayed the split load with sledges. Marshall said of the terrors they faced, 'Crossing crevasses all rest of afternoon, a terribly treacherous surface, as there is nothing to indicate their presence . . . Shackleton and I both went through just before camping. This has upset Shackleton and he is not in good form. A whole line of nasty stuff ahead of us for tomorrow.'

After facing several dices with death, and his nerves shot, Shackleton wrote that 'every step was a venture'. Wild added, 'I would rather walk 40 miles than do it again.' But if Wild believed the worst was over, he was sadly mistaken.

On 7 December, as Wild led Socks at the rear of the party, some 1,700 feet up the glacier's nursery slope the ground opened up beneath him. Plummeting into the abyss and certain death, Wild was yanked to a halt. By a sheer stroke of luck, his sledge had spun on its side and jammed itself across the hole, with his harness saving him from falling. The same thing once happened to me on the Beardmore Glacier. As I attempted to cross a crevasse bridge, with my ski sticks strapped to my wrists, the surface collapsed under me. Dangling above an enormous hole, I realized that I was only being prevented from falling to my death by one of my ski sticks, which was wedged above me, and still fastened to my wrist.

Socks was not so fortunate. His weight snapped his trace to the sledge, leaving him to fall into the darkness. The party thanked their lucky stars that at least the sledge was safe, as it was carrying food and two of the four sleeping bags. Had it been lost, they would all have certainly died.

Still, the loss of Socks was another monumental blow. They would now have to drag over 1,000 lbs between them, and they had lost the pony's meat, which could have lasted them for two weeks. It was poor consolation that they could now eat his feed: 40 lbs of maize and 31 lbs of a compressed high-energy concoction of meat, carrots, milk, sugar and raisins. Determined to present a brave face to his men, Shackleton said, 'Difficulties are just things to overcome after all.'

When fearing failure, and sensing the gradual spread of panic, I have discovered that the best way to combat such negative thoughts is to mentally rehearse how I would react in a worst-case scenario. For example, when approaching a crevasse field, I conduct a mental rehearsal of what exactly I will do, step by step, should I happen to plunge into the maw of a 100-foot fissure. Feeling that you are prepared, no matter what, has a tremendously positive effect.

An exhausted Shackleton wrote of the journey, 'Started at 7.40 a.m. on the worst surface possible sharp-edged blue ice filled with crevasses, rising into hills and descending in gullies.

It is a constant strain on us all both to save the sledge from breaking or going down a crevasse and to save ourselves as well: we are a mass of bruises where we have fallen on the sharp ice . . . It has been relay work all day for we could only take on one sledge at a time . . . Still, we have advanced 3 miles to the south . . .'

Each minute seemed an eternity, as anxieties gnawed and frostbitten toes or fingers vied at each step with the sharper pain of underwear tugging at raw groins. As hunger and tiredness pushed them to breaking point, each eyed the other with disdain. Wild snapped at Adams and Marshall for not pulling their weight, writing that Marshall was 'a big hulking lazy hog' and wishing 'he would fall down a crevasse about a thousand feet deep'. He was somewhat kinder to Adams, admitting that he tried his best, although it was still 'a very poor best'. Together, he claimed that Marshall and Adams were 'two grub scoffing beggars'.

Marshall, who never needed any invitation to criticize Shackleton, now found it difficult to bite his tongue. Shackleton might have led bravely from the front, in spite of his snow-blindness, but Marshall had no sympathy, claiming it was his own fault and he was tired of his 'moaning'. In a litany of complaints, Marshall described his cooking as 'useless', and claimed that Shackleton was like 'an old woman, always panicking'. Later, at the end of his tether, he told Shackleton to his face that their journey 'should not have been attempted *from that base* [Cape Royds] for the sole purpose of getting to the Pole, and the mentality of any man who tries to defend this basic decision is warped'. Although Shackleton tried to rise above Marshall's black moods, he couldn't resist landing a few jabs in his letters to Emily, where he described Marshall as 'a bit young in mind [and] inclined to resent discipline'. In being so vocal, even under such testing conditions, Marshall was compromising the safety of the group, and no doubt Shackleton must have wondered if he had been wrong to bring him.

In time, each man chose to traverse the dreamworld of whiteness and weird shapes in silence. Keeping their complaints, pains and fury to themselves, they dragged their sledges through the eerie crevasses. The pole was still 350 miles away, and the only certainty they had was that they were heading south.

The pole might have been their goal, but they also still had scientific work to complete. Using their theodolite and hypsometer, they mapped out the virgin terrain and recorded the heights of many of the peaks around them, naming numerous mountains and glaciers after those with links to the expedition: Dorman, Mill, Keltie, Adams and Wild. In a moment of charity, Shackleton even chose to name one after Marshall, although he received little gratitude for it.

Again, there has been some debate surrounding the naming of some of these landmarks, particularly Mount Donaldson. It has transpired that while travelling to New Zealand Shackleton had struck up a friendship with an Isobel 'Belle' Donaldson. Once more, it is impossible to say if the relationship was in any way romantic, but just before the *Nimrod* set off for Cape Royds, Shackleton wrote her a letter, which he signed 'Your Polar Man'. It seems the 12,900-foot mountain was named after Isobel, with whom he kept in contact for many years afterwards. At this rate, Shackleton was in danger of running out of mountains to name after his many female acquaintances!

Discovering new mountains and glaciers was thrilling in itself, but Wild made the most startling discovery of all: coal. Inspecting the rocks, he found dark seams up to 8 feet thick. This was not only the first time such a valuable resource had been found on the continent, it was a groundbreaking scientific discovery in itself, proving that millions of years before Antarctica had been a warm, swampy environment. If the expedition could already be classed as a success, by virtue of the men having beaten Scott's record, then this discovery would really boost the scientific element. Indeed, it was in the hope of discoveries such as this that the Australian government had listened to David and agreed to give Shackleton such a generous donation.

The mood of the group improved further when the top of the glacier came into sight. Shackleton recognized that the pole was within their grasp. However, if they were to have any chance of success, further sacrifices would have to be made. 'We must march on short food to reach our goal,' he wrote, cutting rations once more.

Setting up another supply depot, they left behind four days' worth of provisions, taking with them biscuits and a small amount of pemmican hoosh mixed with pony maize. They would need to travel at least 22 miles a day, something they had not previously achieved, even on full rations. Shackleton knew this last push was all or nothing, writing, 'We have burnt our boats now.'

The march ahead became particularly unpleasant. In the higher altitude, the air grew thinner and colder, making it more difficult to breathe. The air at 10,000 feet near the pole has the same effect on humans as 13,000 feet on equatorial mountains. Dehydration is another problem, caused by increased respiration and the dry atmosphere; on the plateau, the men also had just half the oxygen they had had down on the Barrier.

Shackleton's awful experience with altitude actually helped me to prepare for the Trans Globe Expedition. Knowing the difficulties he had faced, when it came to me choosing a winter base camp, I selected a site beneath Ryvingen Mountain, on the basis that it was 5,000 feet above sea level. Using this site helped us to acclimatize to high altitude over seven months before we attempted the main crossing. Shackleton was of course in uncharted territory so had no idea he would have to prepare for such an obstacle.

With Marshall noting that man-hauling at altitude was 'killing work', he now kept his eyes on Shackleton. He knew these were just the sort of conditions that could prove fatal for a man who had previously suffered breathing difficulties. For now, at least, Shackleton soldiered on, without any outward indications of imminent trouble, but still Marshall wrote, 'Shackleton rather done . . . won't stand a higher altitude.'

On they pushed, finally reaching the 86th parallel on Christmas Day, at which Shackleton decided it was time to reward his men. Extra pemmican was subsequently served for breakfast, a slither of cheese added to the regular lunch of four biscuits and tea, and all the stops were pulled out for the evening festivities.

Crammed into a two-man tent, the four men dined on a concoction of pony food mixed with Oxo, pemmican and biscuit before polishing off a dessert of plum pudding cooked in brandy-flavoured cocoa-water. Cigars were then handed around, which were clenched between frost-split, sun-blistered lips, soon filling the tent with their pungent smell. Suffice to say, while the thermometer showed 52 degrees of frost and a gale was blowing outside, this was a time of sheer bliss. Wild felt particularly cheered: 'For the first time for many days I feel replete and therefore will not make any nasty remarks about anyone, although I should very much like to.'

The men woke the next morning with a renewed vigour to

attack the remaining 280 miles to the pole. Noting everyone was in high spirits, Shackleton decided they would now be able to accept the bad news he had to give them. Given their rate of progress, rations would now have to be stretched to last ten days rather than seven. 'We will have one biscuit in the morning, three at midday, and two at night,' Shackleton wrote. 'It is the only thing to do for we must get to the pole, come what may.'

Shackleton was truly putting his ambitions to the test. Reaching the pole was everything to him. In his diary he wrote a verse of 'Love in the Valley' by George Meredith, which illustrated his mindset:

> Hard is our love. Hard to catch and conquer.
> Hard, but oh, the glory of the winning were it won.

Whenever I was really up against it on a tight polar schedule, and still striving onwards, I would imagine that my grandfather, a pioneer in Canada and Africa, and my father and my uncle, both killed in the world wars before I was born, were alongside me, willing me on. This was always a tremendous comfort and gave me a renewed sense of purpose. I imagine that at a time like this Shackleton thought of Emily and his children, and how each step towards the pole was another step towards making them proud of him.

Soon after, all non-essential equipment was discarded to lighten the load: spare sledge runners, scissors and clothes all left on the ice. These sacrifices allowed the men to get closer to their target schedule, passing the mountain range by the end of December and finally setting foot on the flat surface that would lead them to the pole. However, still more sacrifices had to be made if they were to make up 15 miles a day.

A tent was now discarded, and another food depot established. It would be tight but, with this, Shackleton estimated that, with good fortune, they could reach the pole by 12 January and still have enough time to get back to Hut Point by 28 February, the cut-off date before the *Nimrod* set sail.

In comparison to what had already been achieved, this last push sounded possible. After weeks of torture, they were just a few days away from all they had set out to achieve. And yet the men could now barely walk, even with the reduced weight they were carrying. Each new ache of muscle or tendon became an obsession as blisters and chafing sores screamed their presence. To make their meagre rations appear bigger they dipped their biscuits in their hot tea so they swelled. Alas, it did not fill their stomachs any more. They were surrounded by ice and snow, but fuel rations had been cut so low they could no longer afford to thaw out ice for water. Dehydration was beginning to bite hard. Their sweat-sodden clothes froze to their skin, making them ever colder, but they had dumped their spare clothes, so had nothing to change into. When Marshall took the men's body temperatures, he found that they were all on the verge of hypothermia, hovering just above 95 degrees Fahrenheit.

I thought this sounded extreme, so I checked it out with Dr Mike Stroud, who said: 'This is not something I have come across before, and I am very sceptical. You only have to think about it for a moment to realize that to "monitor body temperature closely" is not possible on [such] a polar expedition. Accurate measures would have to be rectal, as mouth temperature is affected by air temperature, and when the air is very cold the readings will appear to be lower than reality. Added to that, a drop of only 2 degrees (presumably Fahrenheit, but even Centigrade) isn't enough to affect one's thinking. It therefore seems more likely that it was a low-blood-sugar problem, similar to that which I suffered a few days from the Pole.'

Whatever the source of their issues, the men collapsed

regularly and had to be helped to their feet, while frostbite remained a real concern. On the same route in the 1990s, four of my toes froze. My co-man-hauler on that trip, Mike Stroud, wrote in his diary, 'Ran could no longer bear the torment without reaction. The swelling on the right foot was now punctured by a deep hole, red, angry, inflamed with pus, and the black and swollen toes of both feet were demarcated by a vivid red line at the base where good tissue fought against the infection. The stench of rotting flesh was added to our already evil body odours.' If this Antarctic experience taught me a new meaning for the word 'cold', it did likewise with 'pain'. Broken bones and teeth, torn-off digits, frostbite and chronic kidney stones had seemed unpleasant, but during those nights in Antarctica I knew real pain for the first time.

Gritting his teeth through the unrelenting torture, Shackleton wrote, 'It is hard to know what is man's limit,' still urging his men forward. Despite it all, the men man-hauled 12.5 miles on New Year's Eve, an incredible achievement. 'Please God the weather will be fine during the next 14 days,' Shackleton prayed. 'Then all will be well.' Even Marshall could not fail to admire Shackleton's super-human willpower, which drove on his team, with the expedition leader attacking each step with every ounce of gristle at his disposal, finding the strength to continue to lead from the front.

The advent of the New Year brought further encouraging advances of 11 miles, attaining a world record in the process by beating the furthest distance ever travelled towards either Pole (87 degrees 6½' north, claimed in 1906 by Robert E. Peary). Shackleton and his men had also surpassed the highest latitude any human being had ever reached.

However, soon after, Wild, who had always been loyal to Shackleton, felt that they could go no further, writing, 'Flesh and blood would stand no more.' Marshall also found that all the men's temperatures had dropped below 94 degrees, so low that the thermometer couldn't pick them up. Privately, Shackleton began to think for the first time that they might have to turn back.

On 2 January 1909, after a punishing 10-mile march, with the pole at least ten days away, Shackleton wrote in his diary, 'We are not travelling fast enough to make our food spin out and get back to our depot in time.' The crunch was fast approaching. They could reach the pole, but the prospect of them all getting back alive would be slim, and then who would know of all they had achieved? This was clearly running through Shackleton's mind as he grappled with his decision. 'I feel that if we go on too far it will be impossible to get back over this surface,' he wrote, 'and then all the results will be lost to the world.'

If he had been operating alone, Shackleton might very well have risked his life to achieve the impossible. Yet he had to take into account the well-being of Adams, Marshall and Wild, who would follow him to the end of the Earth if he told them it could be done. 'I must look at the matter sensibly and consider the lives of those who are with me,' he confided to his diary. 'Man can only do his best and we have arrayed against us the strongest forces of nature.'

On 6 January, Shackleton admitted defeat. They had reached 88 degrees 5', some 115 miles from the pole, but he could not risk the health of his men any longer. 'I would fail to explain my feelings if I tried to write them down, now that the end has come,' he wrote. The pole might have been beyond reach, but with the weather good and the surface flat, Shackleton rallied his troops for one final dash south, to add to their already incredible record. His goal was to get within 100 miles of the pole, which with one push, and a little creativity, was doable.

To date, Shackleton had measured distance travelled via statute miles. Getting to within 100 statute miles of the pole would now be impossible, yet he realized that they might be able to get below 100 nautical miles. With this in mind, they made one last depot, marked only with a bamboo pole and an empty bag as a flag, and set off.

This was a significant risk. The folly of the decision was exposed soon after when a blizzard confined the men to their tent for forty-eight hours, forcing them to eat their dwindling supply of rations. Lying in their bags, crammed together like sardines, their frozen breath swirling about, each man must have privately doubted whether he had made the right choice. Wild wrote, 'I think today has been the worst we have yet experienced. We have found it utterly impossible to keep ourselves warm, and we have all been frostbitten . . . we had to camp at 4.30 or I really believe we should have collapsed.'

Finally, on the ninth, the blizzard died away, but still they did not turn back. Leaving behind the tent and sledge, they set off with just a few biscuits and pieces of chocolate, as well as a brass cylinder containing their King Edward VII Land stamps and Queen Alexandra's silk Union Jack flag. While the weather was good, by 9 a.m. Shackleton had no choice but to call a stop. 'We have shot our bolt,' he wrote. No one had anything more to give and limitless whiteness still stretched to the horizon. But were they within 100 nautical miles of the pole, as Shackleton had hoped? It fell to Marshall to check.

Throughout the expedition, Marshall, the navigator, had been using a theodolite to measure the distance travelled by taking sightings from the sun's position at noon. If he were unable to see the sun, he relied on the distance indicated by the sledge meter, which measured their daily mileage. However, for this final push the theodolite had been left behind to save weight, as had the sledge, making it very difficult to state with certainty their exact position. Marshall therefore used guesswork to announce they had reached 88 degrees 23' S, 97 nautical miles from the pole. Some have disputed this final reading, helped by Marshall, who later said that 'the facts will be disclosed in a sealed statement, before or after my death'. Despite this, nothing more was ever said about the matter.

The emaciated men took a moment to mark their 'furthest south record' by holding Queen Alexandra's flag aloft for photographs. They had beaten Scott's record by 412 miles. It was an awe-inspiring achievement, certainly when the chaos of the preparation and severe lack of funds and support are factored in.

Yet Shackleton could not enjoy the moment. He estimated that they would have reached the pole had they just carried an extra 25 lbs of biscuits and 30 lbs of pemmican, or if Socks had not fallen to his death. This is probably true. Rather than any argument about skis, dogs or ponies, as I have found myself, it is rations that are the most vital part of the journey south, and they had fallen just short. Nevertheless, Shackleton could not deny that, in the circumstances, they could have given no more. 'Whatever regrets may be, we have done our best,' he wrote.

The men knew just how much missing out on the pole hurt Shackleton, who continued to stare south, dreaming of what lay just beyond. Wild understood that he had put the safety of the men above his own dreams. 'I am perfectly certain that had Shackleton only himself to consider,' he wrote in his memoirs, 'he would have gone on and planted the flag at the Pole itself.'

Shackleton's decision to turn back when he did showed remarkable courage and intelligence. It is for reasons such as this that he continues to be so revered. This was of little consolation to him at this point. He was already thinking of what he would change when he next returned, determined to reach the pole one way or another. But would someone have already reached it by then? And he would have shown them the way! It was too cruel a prospect to consider.

Before he could really digest all of this, he first had to return safely to the *Nimrod*, which was over 750 miles away as the crow flies. Five years before, Shackleton had almost died on the return to the *Discovery*, facing a much shorter distance. With the men's health already deteriorating and extreme hunger setting in, it was vital they found all the food depots on their return march.

22

After easily making it back to the nearest depot on 11 January, Shackleton realized it would be a daunting task to reach the next one in time. It was over 200 miles away and the men had only ten days' worth of provisions, at half rations, to sustain them. Moreover, it had previously taken them seventeen days to walk the same route. In short, the prospects appeared bleak.

Shackleton was also in a bad way. His feet were now badly frostbitten, which Wild noted with concern: 'Both his heels are split in four or five places, his legs are bruised and chafed and today he has had a violent headache through falls, and yet he gets along as well as anyone.' Every step was agony, with the prospect of marching over 14 miles a day to make it to the depot seeming ever more fanciful. However, Shackleton had a cunning solution to increase their speed while also saving his feet.

The journey south had been marred by marching into the bitter wind. This time the wind was behind them, and Shackleton intended to use it. Remembering the attempt to utilize the floor cloth of their tent as a sail while with Scott, he now did the same, with the wind propelling the men through the frozen glacier at an incredible pace. Travelling up to 20 miles a day, they were exerting nowhere near as much effort. All being well, this moment of inspiration would see them reach the next depot in good time.

Yet finding the depot would be a challenge in itself. As had been the case on the return march with Scott, locating such a tiny location without the aid of detailed maps or modern inventions such as GPS was dependent on a large slice of luck. As Shackleton wrote, 'It has been a big risk leaving our food on the great white plain, with only our sledge tracks to guide us back.'

Their outward trail was initially still visible for them to follow but as they marched further their prints faded away. They then realized that the sledge meter had fallen off. This would make it difficult to work out how far they had travelled each day, and to work out how far it was to the depot, as well as to Cape Royds. 'This is a serious loss to us,' Shackleton admitted.

Their only hope now was to pick up familiar landmarks. This would tell them they were at least heading in the right direction and they could also estimate how far they needed to travel. Thankfully, for now at least, the mountain range allowed them to pick out certain landmarks, such as Mount Buckley, and with the wind enabling them to travel 29 miles on 19 January, they picked out the depot in good time.

The men wasted no time in gorging themselves on food. Yet Shackleton, who looked more in need of it than most, couldn't stand the sight of it, having developed an upset stomach. Setting off the next day, the state of Shackleton's health became ever more alarming. Marshall, who had begrudgingly come to admire the leader's tenacity, watched him nervously. 'Pulse on march thin and irregular at about 120,' he recorded after one examination, speculating that if it got much worse, they would have to carry Shackleton on the sledge, just as Scott and Wilson had done six years before. Shackleton couldn't stand that humiliation again, but he was soon so weak he could no longer haul. For now, he had no option but to let the others pull while he staggered alongside. Rather than recording Shackleton's illness as a weakness, Wild chose to put it down to his incredible strength, writing, 'for a good six weeks he has been doing far more than his share of work'.

Shackleton drew ever more on his deep reserves, refusing to be blamed for holding the party up, the bitter memory of his time with Scott propelling him on. The wind continued to help. Even while navigating the hazardous crevasses, they were able to achieve 15 to 20 miles a day. Shackleton's stomach also settled, so he could at last hold down some food, giving him a new lease of energy. Although cold and frostbite continued to hit him hard, he was also able to breathe more easily as they made their way to a lower altitude. For Shackleton in particular, this was crucial.

The next depot was 40 miles away but they had almost run out of rations. 'We look at each other as we eat our scanty meals,' Shackleton wrote, light-headed and riddled by migraines, 'and feel a distinct grievance if one man manages to make his hoosh last longer than the rest of us.' Having devoured the last scraps of pemmican and pony maize, all that was left was Marshall's 'Forced March' tablets, which consisted of cocaine and caffeine. This was not enough. Each man in turn collapsed in the snow before somehow finding some energy to stagger back to his feet. Hollow-eyed and skeletal, they were now barely alive, their legs aching, their shoulders raw from the weeks of sledge traces biting into their skin and the tips of their noses red raw.

When Adams fell, mumbling and incoherent, and could not return to his feet, it was apparent that the men were not going to reach the next depot. If matters weren't desperate before, death now lingered in the shadows.

Marshall, who had been accused by Wild of not pulling his weight, and whose dark mood had permeated much of the march, now came into his own. Spotting the distant outline of the rock formation where the depot was situated, he understood that he alone still had the strength to reach it. Bravely volunteering to go solo and return with food, he set off, knowing it was make or break for all of them.

Staggering between a range of crevasses, Marshall fell often, and was in constant danger of dropping into one, but he always dragged himself to his feet, shuffling or even crawling, refusing to let the men down.

On arrival, the ravenous and exhausted Marshall somehow resisted the temptation to eat his rations, let alone feast on anyone else's. To his enormous credit, he knocked back only two lumps of sugar for energy, gathered as much food as he could carry, and made his way back to the men, aware that every minute was vital.

By the time he reached them, the men had not eaten for thirty-six hours and were in a very sorry condition. Eagerly taking the supplies from Marshall, Wild lit the stove and made a steaming pot of hoosh and horsemeat, washed down with a mug of tea, with biscuits for dessert. 'Good God how we did enjoy it,' he wrote. It was the 'finest meal we had ever tasted'. Shackleton counted his blessings that the men had enjoyed a lucky escape, and all thanks to Marshall. 'I cannot describe adequately the mental and physical strain of the last 48 hours,' he wrote of the worst period of the expedition by far.

Things improved a little the next day; they had food in their stomachs, had left the hazardous glacier behind and were back on the smooth Barrier. It was now only 50 miles to the next depot, and they had enough food to last six days. In contrast to what they had just endured, this was truly luxury.

However, while in principle this seemed easily achievable, this was without factoring in the deteriorating health of the men. Shackleton and Adams had suffered badly, and it was now Wild's turn. Having developed a crippling bout of dysentery, which he put down to eating pony flesh, he frequently had to stop due to diarrhoea. This caused further dehydration, leading him to suck desperately at ice or snow for some relief. Utterly exhausted, he was barely able to pick up his sledge trace to tow the sledge, let alone put one foot in front of the other for miles on end. Marshall gave him some medicine to keep him going but it only made him drowsy. Now nothing but skin and bone, and surviving on the only thing he could hold down, four biscuits a day, he looked close to death and wanted nothing more than to crawl into the tent at the end of the day and die.

Recognizing Wild's rapidly deteriorating state, Shackleton forced his friend to accept one of his cherished biscuits. It was all Wild could eat, and Shackleton knew it could make the difference between life and death. It was a tremendous gesture, when each morsel of food was vital to each man and a biscuit was considered a real treat. Afterwards, Wild would proclaim his devotion to Shackleton for the rest of his life. 'I do not suppose that anyone else in the world can thoroughly realise how much generosity and sympathy was shown by this,' Wild wrote. 'By GOD I shall never forget it. Thousands of pounds would not have bought that one biscuit.'

Selfless gesture or not, biscuits alone could not sustain Wild. He had been unable to eat anything else for over six days, and all the while his body was continually being flushed of any remaining nutrients. 'If I don't soon get over it, I am afraid I shall have to be left on the Barrier,' he shared in his diary.

They had enough food to last them to the next depot, but the men all soon developed diarrhoea and faced the same challenges as Wild. On 4 February, Shackleton recorded, 'The entire party collapsed and was unable to move.' Adams had to leave the tent seven times during the night to relieve himself while Marshall said the camp was 'like a battlefield ... Outlook serious.' Yet the men could not afford to sit idle. They had no choice but to battle on, their only solace that the wind was enabling them still to cover around 20 miles a day. Yet Cape Royds was still 240 miles away.

Thankfully, with the diarrhoea subsiding, and the men able to hold down their food, the next depot offered a rare treat: the remains of Chinaman. Reaching the depot on 13 February, the men ate all they could, including the congealed blood, which Wild made into a 'beautiful soup'. A further treat came two days later as Shackleton celebrated his thirty-fifth birthday. Extra hoosh was served to mark the occasion, while the men presented Shackleton with a truly heavenly gift; a cigarette, made of tobacco shreds from the men's bags, lovingly packed into a paper tube. 'It was delicious,' Shackleton happily reminisced.

To date, the only success of the return march had been the tremendous distances they had covered thanks to harnessing the wind. Without it, there was next to no chance of them reaching each depot in time. Now passing familiar landmarks, such as Mount Discovery, their pace slowed to around 12 miles a day, again stretching their rations. 'We are appallingly hungry,' Shackleton complained, now down to consuming just 2,000 calories a day, less than half what was required.

Reaching the next depot, on 20 February, they devoured more hoosh and biscuits, as well as jam and cocoa. To their joy, there was also a supply of cigars from Shackleton's Tabard tobacco company, which helped keep them warm. But their rations would not last the 140 miles still needed to reach Cape Royds. If they were careful, and made good speed, they had enough to last four days, which might allow them to reach Minna Bluff, 70 miles away, where Joyce was supposed to have prepared another depot. This of course depended on whether Joyce had followed Shackleton's instructions. And even then, they would still have to find it.

On 21 February, a snowstorm threatened to confine the men to their tent, but they had no choice but to keep going. Such was the wind's ferocity, they could barely see in front of them, while temperatures dropped to minus 35 degrees Fahrenheit, the icy wind cutting through to their skin and the howling wind hurling shards of ice into their faces. As the chill permeated their bodies, they faced a constant battle to maintain their core temperature, their clothing all but useless in the face of the repeated katabatic gales. 'It is neck or nothing now,' Shackleton wrote. 'Our food lies ahead and death stalks us from behind.'

Somehow, and in these conditions, they marched over 40 miles in just two days. Even men in prime physical condition would have struggled to match this. Despite being 'so thin that our bones ache as we lie on the hard snow in our sleeping bags,' Shackleton inspired his ailing men to find deeper and deeper reserves within themselves when they appeared to have been exhausted long ago.

To their great relief, on 22 February, fresh tracks appeared in the snow, along with discarded tins and cigarette butts. They were clearly on the right track, but had Joyce left food at the depot? One day's march to Minna Bluff remained but the men had eaten all their food, bar a few biscuits. If they did not reach and locate the depot that day, the game would be up.

Setting off at 6.45 a.m., they searched in vain. By midafternoon there was still no sign of it. The biscuits were now gone and darkness would soon descend, making it virtually impossible to pick out the fluttering black flag that would mark the spot. There remained just a handful of hours to locate it.

Eyes wide, and necks like swivels, they crawled forward, scanning the Barrier in all directions, their hopes hanging by the slenderest of threads. Then, at 4 p.m., Wild saw a light flash in the distance. He couldn't be sure if it was the depot, but it was all they had to go on.

Urging them forward, he found the light was the sun's reflection bouncing off a biscuit tin perched on top of a 10-foot-high mound, marked by three flags fluttering in the wind. They were saved. Joyce had not only made the depot as instructed but had left freshly boiled mutton, plum puddings and eggs. And there was more good news. In a note, Joyce confirmed the arrival of the *Nimrod*, now under the command of Captain Evans of the *Koonya*, following the dismissal of Captain England.

However, it was now 26 February, and the ship was set to return to New Zealand on 1 March. An almighty final effort was needed to get back in time. If the gods smiled on them, they should just about make it, but they could not afford any delays.

The gods, sadly, weren't listening to their prayers as, once again, diarrhoea struck, slowing the march and leaving some paralysed by stomach cramps. Sinking to their knees in the snow, they knew that they had to keep marching, but then a ferocious blizzard left them with no option but to seek shelter in their tent. Time was now ticking away.

On 27 February the blizzard blew out, allowing them to push on, determined to make up the lost time. Incredibly, they did just that, marching over 26 miles, even allowing for Marshall's constant outbreaks of diarrhoea, until he could take it no more.

Setting up camp, Shackleton estimated that they still had 30 miles to travel. The *Nimrod* was set to sail in just thirty-six hours. Yet it was clear that Marshall had nothing more to give.

By now he was so malnourished his bones were in danger of protruding through his wafer-thin skin and the muscles on his once-bulging legs had wasted away to nothing. If he were to continue in this state, they would never make it in time. Shackleton now had to make a bold decision.

Adams was thereby instructed to stay behind to tend to Marshall at camp, while Shackleton and Wild would set off for Hut Point, the nearest refuge. Shackleton had instructed some men to stay there, but even if it was unmanned, Shackleton had also specified that nearby Observation Hill should be manned, with signals sent at regular intervals. At the very least, they could flash a signal that they were safe and well and communicate that the *Nimrod* should not leave without them.

For the final march they carried only two sleeping bags and a small amount of food. They just had to 'trust in providence' that a blizzard didn't strike, slowing their progress and leaving them exposed to the elements.

Setting off at 4.30 p.m., they marched through the night and into early morning before they were sure they were in sight of Observation Hill. Frantically, Shackleton flashed a heliograph, but to his frustration, they received no signal in return. Had the *Nimrod* already set sail?

Having no choice but to march on, with all their food now eaten, Shackleton and Wild bordered on the delirious. At one point, they raced excitedly towards a group of figures in the distance, believing it to be a search party from the *Nimrod*, only to find it was a family of penguins. Midday came and went, meaning they had been marching for over twenty hours. Shackleton was aware that Wild was becoming discouraged. 'Frank, old man,' he said, grabbing his friend's arm, 'it's the old dog for the hard road every time.' Wild mustered a small grin; while the younger men in the group were laid up in the tent, the two thirty-five-year-olds had somehow kept on going.

With the end of the Barrier in sight, the men reached the sea ice, hoping to find a path to Hut Point. However, the ice was so unstable there was no way across. They had no choice but to take the longer land route around. Even this was not safe, as Shackleton and Wild well remembered, for it was nearing where George Vince had plunged to his death during the *Discovery* expedition.

Cautiously navigating their way over the uneven ground, poor visibility slowing their progress ever more, the men reached the top of the slope, hoping to be able to see the *Nimrod* in the distance. But she was nowhere to be seen. Nor was there any sign of anyone at Hut Point. This was a shattering blow. Wild recalled that they were both 'beyond speech' as they began to realize that, at best, they would be stuck in Antarctica for at least another winter. They had to hope that, at the very least, food had been left behind for them.

On entering the deserted hut they found a note from Professor David. All was not lost. The *Nimrod* was no longer at Cape Royds, but between Hut Point and Cape Royds. This explained why they had not been able to see her. However, the note also said that the ship would be there only until 26 February. It was now the night of 28 February. There was every chance she might have gone already. Shackleton was furious. He had left clear instructions that the *Nimrod* was to depart on 1 March.

Shackleton and Wild tried to send a signal to where they thought the *Nimrod* might be located, if, that is, she had not set sail to avoid entrapment. Igniting a carbide flare with urine, they waved it in the boat's rough direction, hoping it would be seen. In the meantime, all they could do now was to wait and pray. If the *Nimrod* didn't appear within the next few hours, they would be stuck and would soon have to return to the Barrier to save Marshall and Adams, carrying with them this depressing news.

To their great joy, and surprise, the *Nimrod* soon steamed into view, the flare having been spotted from the masthead by the one-eyed Aeneas Mackintosh. Shackleton and Wild hugged each other deliriously. They had not slept in over thirty-six hours, but their efforts had not been in vain. Wild wrote of seeing the ship, 'No happier sight ever met the eyes of man.'

Less than two hours later, the bedraggled, skeletal pair boarded the *Nimrod*, to the delighted shock of the crew. It had been assumed that they had died, and the sight of them left some believing they were certainly not far from it. Concerned for their health, the men hurriedly prepared bacon and fried bread, which Shackleton and Wild wolfed down.

Such was Wild's condition he had to be sent to bed, but Shackleton gathered a small party consisting of Mackay, Mawson and Michael McGillion, who was from the new crew of the *Nimrod*, and returned to the Barrier to save Marshall and Adams. Recognizing that by now their food would be extremely low, if not have run out altogether, Shackleton decreed that the party could not rest until they reached them. So, began yet another hellish march, this time for twenty-four hours, with not a wink of sleep for the exhausted Shackleton, who had been marching practically non-stop for over two days and nights.

Finally, by midday on 2 March, Shackleton caught sight of the tent. Inside, Marshall and Adams were in a pitiful state, with no food to spare. To their joy, Shackleton had come laden with provisions, as well as the great news that they had caught the *Nimrod* just in time. Soon they would be on board and returning home. It was news to lift all of their spirits.

After another twenty-four-hour march, everyone was safe and on board the *Nimrod*. Shackleton, who had lost 24 lbs on the polar journey, could now rest, and stuffed himself on anything edible. In 1993, after ninety-four days of Antarctic man-hauling, I lost 55 lbs. While I soon put it back on, I often wonder what strain that might have put on my body, not least my heart, as in 2003 I suffered a massive heart attack at an airport and survived only thanks to a mobile defibrillator being immediately available. Would Shackleton have also suffered further damage to his heart after such an excursion? After all, it is well known that sudden weight loss can damage the blood vessels, leading to fluctuations in heart rate, and increased blood pressure, all of which increases the risk of heart failure. Shackleton might have been back to safety at last, but in my mind there is little doubt that he was now storing up troubles for the years to come.

At this moment, such matters did not concern Shackleton. He might not have achieved his ultimate goal of reaching the pole, but he had certainly carved a name for himself in history. He had travelled further south than any other human being, charted an area of unknown terrain, and discovered coal. This would surely cement his reputation as an explorer and a scientist, while his leadership skills had been proven to be of the highest calibre. While he had been on the ice his team had also completed various valuable research tasks and they had of course also scaled Mount Erebus.

Somehow, Shackleton had urged and cajoled his team to man-haul over 1,755 miles in the most treacherous conditions, all suffering beyond what most human beings ever have to endure. Despite experiencing extreme hunger, cold and exhaustion, they had all come out of this hell alive. At every juncture, Shackleton had adequately weighed up the risk, pushing the men to their absolute limit but never willingly putting their lives at stake.

Perhaps more than anything it was his decision to turn back when less than 100 nautical miles from the pole that was to earn the most admiration. Everyone was well aware that this was Shackleton's ultimate ambition and he had made unbelievable sacrifices to get so close. It must have been an agonizing decision. Many would have chosen to go for broke, come what may. But despite his own personal ambitions, Shackleton refused to put the safety of his men in question. As Adams himself later said, if they had but marched just one hour more, they would have surely died.

It is a mark of the loyalty he inspired that when Shackleton was suffering with dysentery, starvation and exhaustion, he had turned to Wild and asked if he would accompany him again in the future, once more trying to reach the pole. Despite all they had been through, and still had to face, Wild immediately said yes.

For now, though, any thought of another expedition would have to wait. First, they still had to escape the ice, which now threatened to trap the ship.

23

Frantic arrangements were made to set sail for New Zealand as soon as possible. It was now 4 March and ice was already building around the *Nimrod.* All necessary equipment and provisions from Hut Point were quickly loaded on board, with anything that was not vital left behind. This included twelve months' worth of rations that had been earmarked to sustain the party for another winter on the ice. No matter. They would be put to good use by any future expeditions. For now, they had to leave McMurdo Sound before it was too late.

Fierce winds crashed newly formed ice against the fully loaded ship as the *Nimrod* raised anchor and set sail, with the men singing 'Auld Lang Syne' in celebration. Shackleton, meanwhile, looked back at the hut as it slowly disappeared from view, 'with feelings almost of sadness'. His great adventure was over, and in his heart he felt he had failed. On telling the men about their tremendous achievements, there was always the caveat, 'all this is not the Pole'. Back home, he would paint a picture of triumph, and he had indeed set a new furthest south record, but he knew his detractors would revel in his 'failure'.

It was perhaps with this in mind that he decided against heading straight for Lyttelton. He instead ordered Evans to head to an unexplored region just west of Cape Adare, desperate to salvage another new discovery to return home with. It was almost a risk too far. Impending ice looked to trap the ship and, not wishing to push his luck any further, Shackleton abandoned this final mission, with the *Nimrod* leaving the Antarctic standard ice zone behind on 10 March.

It was not to matter. When Shackleton swapped stories with the teams who had ventured on to the ice, to fulfil the various missions he had set them, he found that some had made historic new discoveries. Although to do so they had to put their lives on the line.

In Shackleton's absence, the Western Mountains Group of Priestley, Brocklehurst and Armytage had been tasked with conducting geological studies over the sea ice on the far western side of McMurdo Sound. Having set up camp one night, Armytage found to his horror that the floe they were on had broken away from the coastline and was heading out to sea. It was only by sheer good fortune that the floe made brief contact four hours later with a zone of fast ice, allowing them to get their sledges, gear and themselves to safety. 'We had only just got off it,' Armytage wrote, 'when the floe moved away again, and this time it went north to the open sea.' Of particular concern to Brocklehurst had been the circling whales: 'The killers were all around the foot of the glacier, great ugly brutes deprived of their unusual breakfast.' Priestley said of their near-miss, 'May I never have such an experience again.'

David, Mackay and Mawson, who had set off to locate the South Magnetic Pole on 5 October 1908, had endured their own fair share of drama. In one of the longest feats of man-hauling ever achieved, this group had travelled 1,260 miles over 122 days, encountering horrific weather conditions and hazardous terrain, not least on the Drygalski Ice Tongue. Having to relay their heavy sledges, they sometimes travelled only 4 miles a day, while also suffering a shortage of rations and severe hunger.

Although the party was made up of mild-mannered

scientists, the conditions caused tension in the group. Mawson was particularly scathing of Professor David, calling his one-time tutor 'half demented' and threatening to relieve him of command due to concerns that he had gone 'insane'.

Despite all of their troubles, on 16 January 1909, after three months of man-hauling, they located the Magnetic South Pole at 72 degrees 15' south, 155 degrees 16' east. Shackleton was delighted with this news. His furthest south record, coupled with this genuine historic discovery, meant that no one could accuse the *Nimrod* expedition of being a failure. Now that he had proved he could lead a mostly successful expedition, it would also be easier to raise money for another crack at the pole. He was, nonetheless, dismayed to learn just how close he had come to being left stranded on the ice.

As per his instructions, the *Nimrod* had been put under the command of Captain Frederick Pryce Evans, with Aeneas Mackintosh serving as a sub-officer. On 23 January the *Nimrod* had returned to Cape Royds, and all appeared to be going to plan. However, when David and his Magnetic Pole party had not returned by 1 February, Evans had refused to use the ship to search for them, not wanting to use up precious coal. It was for this reason that, soon after, he had also given Shackleton up for dead. Evans therefore disregarded Shackleton's orders that a heliograph signal should be sent every day from Observation Hill and recalled all the men, and supplies, from Hut Point so they were ready to set sail. It was only thanks to Murray, whom Shackleton had left in charge, standing his ground, that the ship waited a little longer. If Shackleton and Wild had not been able to signal the *Nimrod* when they had, not only would they have had to spend another winter on the ice, they might have been trapped there permanently. For if Evans had returned to New Zealand and told the authorities that he believed Shackleton was dead, there was no guarantee that any rescue ship would have been sent.

This must have been chilling to hear, but Shackleton was far too relieved to be angry. Everyone was alive and well, although somewhat battered and bruised. He might also have recognized that no man had acted perfectly in such testing conditions. Each had moments which they might now look back on with regret. For Shackleton, it was the nagging feeling that just a few more rations would have seen him reach the pole.

No matter his own private thoughts about the success of the expedition, Shackleton knew that it was vital that he controlled the story, framing all of their deeds as complete triumphs. A master at PR, at the first available chance Shackleton briefly rushed ashore on the remote and scenic Stewart Island and sent a 2,500-word report to the *Daily Mail*, portraying himself as a buccaneering hero.

News of Shackleton's triumph reverberated around the globe and hit the polar establishment in London like a rocket. Shackleton had seemingly proved all of his doubters wrong, and now they scrambled to cover themselves. While some chose to publicly congratulate Shackleton on his achievement, others were still too bitter to accept his heroic version of the venture.

Unsurprisingly, Markham questioned Shackleton's polar record readings. Writing to the equally sceptical Keltie, he scowled; 'I do not quite see how it [the record latitude of 88 degrees 23' S] is possible,' and to Leonard Darwin (son of Charles), the new President of the Royal Geographical Society, he wrote, 'I cannot accept the latitudes. For 88.20 they must have gone, dragging a sledge and on half rations, at a rate of 14 miles a day in a straight line, up a steep incline 9000 feet above the sea, for 20 days. I do not believe it.'

However, when Shackleton's records were inspected by E. A. Reeves, the RGS map curator, he concluded that the readings were accurate. 'Taking all the circumstances into consideration,' he said, 'I think Mr. Shackleton's latitudes may be accepted as satisfactory.'

Markham was unable to accept this graciously and felt that Scott was still worthy of higher acclaim. He told anyone who would listen that 'Shackleton owed everything to Captain Scott,' and 'Without Captain Scott [Shackleton] would never have been heard of.' Furthermore, despite Shackleton having shattered the furthest south record, Markham chose to criticize his 'faulty management', saying it was to blame for him failing to reach the pole. In his notes about Shackleton he wrote, 'Son of a doctor . . . but Irish . . . He was taken in Scott's southern journey, but broke down and endangered the lives of others. Scott and Wilson saved his life.' In another letter to Darwin, he scowled, 'Shackleton's failure to reach the South Pole when it could have been done by another, and is really a matter of calculation, rather aggravates me.'

In an act of incredible spite, the RGS decided against awarding Shackleton its Patron's Medal, which had previously been awarded to Amundsen, Nansen and Scott for their expeditions. Markham wrote, 'To put him on a par with Captain Scott, or his people on a par with the Society's expedition [*Discovery*], would be a serious mistake in my opinion . . . indeed an outrage.' Instead, Colonel George Milo Talbot was awarded the RGS 1909 top medal for conducting geographical surveys in Afghanistan and Sudan.

Captain Scott heard about Shackleton's new record while travelling with Tom Crean, who had of course also travelled on the *Discovery*. Scott was still furious that Shackleton had broken his promise to camp at McMurdo Sound, but he was nevertheless delighted at the news coming from New

Zealand. Shackleton might have set a new record, but he had not reached the pole itself. If a merchant seaman could get so close on his second attempt on the ice, then surely a naval officer would be able to go one better. If anything, this news gave Scott added impetus to set off on his second expedition as soon as possible.

The Norwegian explorer Amundsen was at least happy to congratulate Shackleton based on nothing more than the expedition's own clear merits, exclaiming, 'What Nansen is in the North. Shackleton is in the South.' He also told Darwin that Shackleton's exploits were one of the 'finest deeds of exploration' and that '[it] reads like a fairy tale and reveals to me a new world'.

It was perhaps the *Sketch* newspaper that perhaps summed up the feeling of the general public on the matter:

> It is one of the symptoms of this age of nerves and hysteria that we magnify everything, that our boasts are frantic and our scares pitiable, that we call a man who plays well in a football match a hero, and that all successes are triumphs . . . but Lieutenant Shackleton is in that rank of heroes whose names go down to posterity . . . when we are all feeling a little downhearted at seeing our supremacy in sport and in more serious matters slipping away from us, it is a moral tonic to find that in exploration we are still the kings of the world.

Of particular concern to Britain at this time was the continuing aggressive rise of Germany, which was now a serious naval power. Therefore many articles on Shackleton's triumph not only celebrated his success as a Brit but also had a jingoistic tone. For instance, an article in the *Sphere* stated, 'So long as Englishmen are prepared to do this kind of thing . . .

we need not lie awake all night every night dreading the hostile advance of "the boys of the dachshund breed".'

The *Daily Telegraph* reported in a similar vein:

> Let us remember at this moment that in our age, filled with vain babbling about the decadence of the race, he has upheld the old fame of our breed; he has renewed its reputation for physical and mental and moral energy; he has shown that where it exerts itself under fit leadership it is still second to none . . . and at a critical time in the fortunes of all the Britons he has helped to breathe new inspiration and resolve into the British stock throughout the world.

While his achievement was a welcome boost to Britain's self-esteem, thanks to the increased printing and distribution capabilities of the new popular press, Shackleton's epic journey reached a much wider audience than had Scott's exploits, ensuring widespread fame. As a result of this, when Shackleton entered Lyttelton harbour he was not just a polar hero but one of the most famous names on the planet, and he intended to enjoy every moment of his long-awaited success.

24

The cheers of the raucous crowd told Shackleton that they were entering the port of Lyttelton. On deck, Shackleton beamed with delight as he caught sight of the thousands who had come from all over to welcome the *Nimrod* back. This was what he had always dreamed of, being cheered as a hero, making his mark on history. Thoughts of unpaid bills and strained relationships were for another day. Today was all about celebrating his triumph.

When he stepped on to dry land, the crowd swarmed around him, wanting to shake his hand or pat him on the back. At this moment he was the most popular man in New Zealand, and he was in high demand. The New Zealand Prime Minister threw him a special luncheon and he was inundated with requests to speak. His visit was in stark contrast to his last to Lyttelton, when he had been desperately scrambling around to raise funds, almost begging people to meet with him. 'It seemed as though nothing but happiness could ever enter life again,' he cheerfully wrote of his reception.

For the next few weeks he capitalized on his new fame by embarking on sold-out lecture tours of New Zealand and Australia, delighting the crowds with his showmanship and engaging personality, earning further acclaim by giving the proceeds away to local charities.

With the opportunities eventually exhausted, Shackleton looked to finally return home. Travelling via various steamers and trains, he used much of the time to write his book, *The Heart of the Antarctic*, with the help of a New Zealand ghostwriter by the name of Edward Saunders. Shackleton could certainly have written the book himself, as his various writing escapades had proven over the years, but it seems that on this occasion he felt far more comfortable telling the tales to Saunders, who would hastily scribble down notes. This played into Shackleton's real strength of telling a story; as he said himself, 'I can talk much better than I can write.'

On the final leg of his journey he sent a telegram to all the media organizations to say that he would be arriving home at London's Charing Cross Station at 5 p.m. on Monday 14 June. He in fact arrived in England two days earlier, under cover, so that he could rest and enjoy some quiet time with Emily and the children before bedlam broke loose.

It had been almost two years since they had last seen each other, and Shackleton had hoped more than anything he would return triumphant. Yet as Emily later revealed, when he discussed his decision to turn back, he said, 'A live donkey is better than a dead lion.' Emily agreed, happy to at last have her 'live donkey' home with her. Despite what Shackleton had previously said to Wild regarding a new expedition, he now told Emily that there would be no more. 'Never again my beloved will there be such a separation as there has been,' he promised, 'never again will you and I have this long parting that takes so much out of our lives.' It was a promise that Shackleton would of course find impossible to keep.

Yet one promise he did keep was ensuring he arrived at Charing Cross Station exactly as planned. From the back of a carriage festooned with flags, Shackleton stood alongside a bewildered Emily and his two young children. Everywhere he turned were people cheering and waving, straining to get a glimpse of the most famous and heralded man in the country. Such was the crush, the gates to the station had to be closed to prevent any more people coming in.

Waving vigorously to the thousands of well-wishers, Shackleton turned to Emily and finally felt a deep sense of satisfaction. He had always felt unworthy of her, but now he felt he had kept his promise. He was now a great man, known far and wide, and sure to make his fortune. While Emily was proud, she was somewhat overwhelmed by the occasion and chose to stay in the background, as she always would. Later, she would say, 'I have never sought publicity, never desired that the world should know of the share I had in his life.'

Among the crowd, Shackleton saw his father and sisters. Kathleen later said, 'He is and always has been my hero – it's rather nice isn't it that the man you like best in the whole world is your brother!' The small boy who had once told her tall tales of his heroism was now a hero in real life.

Casting a shadow over the proceedings were members of the Royal Geographical Society, including Darwin and the scowling Markham. They were there out of duty, rather than to offer any heartfelt congratulations. A far classier response was made by Scott, who Shackleton spotted in the crowd, graciously applauding. Scott had not been able to make up his mind whether or not it would be appropriate for him to attend, but Mill, a friend to both men, had encouraged Scott to go with him.

Fighting his way through the cheering crowds, Scott was able to grasp Shackleton's hand and shout 'Bravo!' in his ear. They might no longer have been friends, but there was certainly a fair amount of mutual respect between them. Both men knew the trials and tribulations of man-hauling across the Antarctic, and it was not for the faint-hearted. Scott's words must have meant the world to Shackleton. He had previously been portrayed as an 'invalid', but now he felt he had truly proven his accuser wrong.

With the congratulations of Scott ringing in his ears,

Shackleton, along with his wife and children, climbed into an open carriage, and made their way past the cheering crowds that lined their route along the Strand. Waving back with a broad smile on his face, Mill said of his old friend, 'I never saw anyone enjoy success with such gusto.'

Following this tremendous greeting, Shackleton enjoyed weeks of congratulations and dinners. On 28 June, he was given a welcome-home reception at the Albert Hall by Prince George, with further royals, politicians and dignitaries among the 8,000 in attendance. Not wanting to hog the limelight, Shackleton ensured that the entire crew of the *Nimrod* were invited and applauded for their efforts. 'The expedition would never have been the success that it was,' he informed the audience, 'if it had not been for the loyal co-operation for the denial of self and for the absolute interest in the objects of the expedition which was shown by the 14 men whom I had the honour to have with me.' In what should have been a moment of celebration, the Royal Geographical Society could not, however, help but engage in a further act of pettiness.

Having decided against awarding Shackleton the 1909 Patron's Medal, it was suggested that at the very least he should be awarded with a gold medal of some sort at the reception. While Keltie reluctantly agreed to this, in an act of spite he tried to ensure that the medal would be insignificant compared to that awarded to Scott. Others at the RGS clearly felt this was a step too far and persuaded Keltie to make it no smaller (or larger) than Scott's.

The official speech recognizing Shackleton's achievement was given by Captain Scott, whose praise included the words, 'We honour him for the manner in which he organized and prepared his expedition, for the very substantial addition he has made to human knowledge, but most of all because he has shown us a glorious example of British pluck and endurance.' The two gave every appearance of being old expedition colleagues, with no sign of any rancour about Shackleton's use of Hut Point – although that might have not been the case if Shackleton had in fact reached the pole.

The ever critical Markham had happily noted to his colleagues that King Edward VII had not seen it fit to attend the reception, with Shackleton clearly not being seen as worthy. Yet he was forced to choke on his words soon after when Shackleton was knighted in the King's Birthday Honours List. From now on, he would be known as Sir Ernest Shackleton. David, Priestley and Mawson were also all knighted for locating the Magnetic Pole. Shackleton no longer needed the Royal Geographical Society seal of approval to make his way. The name Sir Ernest Shackleton alone could now open many doors. The following year, Shackleton attended Buckingham Palace and was installed as a Commander of the Royal Victorian Order, by King George V, the new King, with Edward VII having passed away on 6 May 1910. Shackleton also went to Balmoral, spending the weekend with the new King and Queen and regaling their guests with tales of his exploits. Shackleton was now not only readily accepted by upper-society echelons but positively welcomed.

Aiming to capitalize on his popularity, Shackleton had appointed Gerald Christy, a renowned lecture agent, to arrange 120 lectures across Europe and America. For many, this was a chance to listen to someone who had been somewhere akin to Mars, and they came from far and wide, packing lecture halls all over.

Christy, who had observed a great number of lectures during his career, said Shackleton possessed a 'descriptive power without parallel'. One moment he would have the audience in fits of laughter, the next they would be perched on the edge of their seat. Mill later recalled his 'deep husky voice rising and falling with the movement of his story and sometimes raised to a rafter-shaking roar'. Fellow polar explorer Sir Hubert Wilkins commented that 'Shackleton's blarney was at times purposely designed and accented because he, more than any other man I know, realized the influence and depended on the influence of the spoken word, no matter the integrity.'

America was a particularly good fit for Shackleton. He was self-made and unafraid of success, and his somewhat brash and confident delivery chimed with the American dream. Welcomed with open arms by President William Taft at the White House, he also received a gold medal from the American Geographical Society (AGS) in New York, presented by Robert Peary, the first man to reach the North Pole, on 6 April 1909. Sixty years later, the AGS was kind enough to award me with a similar honour.

However, while in Germany, Shackleton got a taste of the troubles that were to come. 'I can see a strong anti-English feeling in Germany now,' he wrote to Emily. 'When the picture of the Queen's flag is shown there is a stony silence . . . It is a great strain lecturing in Germany, when one does not feel that the people are really with one.'

Shackleton loved visiting countries that were renowned for producing their own brand of historic explorers. This was especially true of Norway, where he received a particularly generous welcome from Nansen and Amundsen, who helped carry him through the crowded streets on their shoulders. He was celebrated just as if he had reached the pole itself. 'I shall never forget the look on Amundsen's face while Shackleton was speaking,' Emily wrote of the occasion. 'His keen eyes were fixed on him . . . the look of man who saw a vision.' Indeed, Amundsen analysed every aspect of Shackleton's journey, making note of his successes, and his failures. At the time, he was preparing for his own journey to conquer the North, although he was in a quandary, following the news that Robert Peary and Frederick Cook had recently reached the pole. Now, hearing Shackleton talk and knowing that Scott was soon to return to the Antarctic, Amundsen wondered secretly if he could beat his English rival to the South Pole.

Before Shackleton could make any plans to return to the ice, he first needed to clear the mountain of debt left over from the expedition, which seemed to be mounting every day. And he knew that if he didn't do this quickly, despite his new-found success, he would risk humiliation and ruin.

25

Before Shackleton set off for the Antarctic, he had placed the *Nimrod*'s financial affairs in the apparently capable hands of his brother-in-law, Charles Dorman. Following his father's death, Dorman was now in charge of the family law firm and was seen as a safe and reliable person to not only oversee *Nimrod* but also the wellbeing of Emily and the children.

Soon after Shackleton disappeared on to the ice, Dorman realized that he might have taken on too much. On learning that he had to find the sum of £7,000 to cover debts (around £400,000 today), he wrote to Emily, 'These worries added to my own business are more than I ever bargained for.'

Dorman's attempts at raising funds from the government or via public appeal in the media fell on deaf ears. The Chancellor of the Exchequer, Lloyd George, turned his request down flat and the *Daily Mail* refused to run any sort of appeal. Vague promises of money were made by various individuals, but only Elizabeth Dawson-Lambton could be relied upon; pumping in yet another £1,000 (almost £60,000 today). Yet with Emily struggling to make ends meet while her husband was away, Dorman had no choice but to give this £1,000 to his sister.

When the *Nimrod* had returned to Lyttelton after dropping Shackleton at Cape Royds, Dorman was informed that she needed urgent repairs. This wasn't in the budget, but if such work was not seen to at once, then there was every chance the *Nimrod* would be unable to return to the Antarctic to pick up Shackleton and the two other exploratory groups. Such a prospect was unthinkable. Short of any other option, Dorman had had no choice but to turn to friends and family to raise funds. The Dorman estate was, thankfully, relatively wealthy and could afford to make some contribution to cover the most urgent costs, and Kinsey, Shackleton's agent in New Zealand, also pulled some strings.

At least these were all expenses that could be quantified on paper. Dorman had no idea of all the generous monetary promises Shackleton had orally made to others, not least to members of his crew. England and David were both due generous bonuses, notwithstanding England's dismissal, and Shackleton had increased the wages of the likes of Adams and Marshall, as well as offering them further bonuses. There simply wasn't the money in the budget to cover Shackleton's largesse, and it was impossible to say how much was needed to cover all his various promises.

Meanwhile, Shackleton, in New Zealand and knowing full well the enormity of his financial issues, regularly chose to donate the proceeds of his sold-out lecture tour to local charities. 'I do not think we could find out all he gave for charity,' Emily once said. 'He did it so often, spontaneously and not always recorded.' Indeed, on his return to England, he came up with a brilliant money-making venture: allowing members of the public to board the *Nimrod* for a tour. Over 30,000 flocked to the attraction, each paying 1s (5p), and proceeds came to over £2,000. Unbelievably, and despite his debts, Shackleton gave all the proceeds to local hospitals. 'I have never heard anyone say he was good at business,' Emily was later to remark.

It seems that Shackleton loved the further acclaim and adoration his acts of generosity brought him. He didn't seem to recognize that he was digging himself further into a hole, and bound to further upset many of his debtors, not least

Beardmore, who was still owed £1,000. Shackleton clearly just wanted to be loved, not realizing that these grand gestures were earning him many enemies from the shadows.

Watching from afar, it must have boiled Beardmore's blood to see Shackleton seemingly earning large sums with his lecture tours, yet still not paying his long-outstanding debts. This was one debt that Shackleton was, however, particularly keen to settle, as he did not want to upset Elspeth.

By mid-1910, it seemed that Shackleton's lucrative lecture tours were all but exhausted. Travelling around America delivering the same lecture night after night, in far flung-towns and cities, had taken its toll, and he was now failing to pack the house in many of the more obscure locations. On one such night, in a 4,500-capacity hall, he failed to attract even 100 people to hear him speak, subsequently telling friends, he was 'sick of it all'.

Shackleton, thankfully, wasn't just counting on the proceeds of his now ailing lecture tours to pay his debts. The publication of *The Heart of Antarctica* in November 1909 promised to bring in substantial funds, although perhaps not as much as Shackleton had originally hoped. While he told Emily that the book and lecture proceeds would guarantee, at minimum, '30 or 40 thousand pounds', and could even go as high as £100,000 (£6 million today), the final sum came nowhere close to these Alice in Wonderland figures. Reviews were at least kind, with *The Times* heralding it as 'The book of the season'.

Newspaper exclusives with the likes of the *Daily Mail* occasionally provided a short-term hit of cash and could net as much as £2,000, but these were irregular sources of income and raised only a drop in the ocean compared to what was owed. Shackleton needed to come up with another lucrative income stream, and fast. Soon he believed he had struck gold.

Before going to the Antarctic, as previously mentioned, he had been appointed as a postmaster by the Prime Minister of New Zealand. Shackleton now believed that the 240,000-plus stamp sheets he had had printed could be very sought after by collectors. 'The stamps alone are worth 10 to 20 thousand pounds,' he told Emily, once more trying to persuade her of his business prowess. Sadly, as was the case with most business ventures Shackleton touched, the scheme was a failure. For whatever reason, collectors were simply not interested and most of the stamps remained unsold. Today, however, the stamps are considered to be highly collectable and valuable.

Things were getting desperate. It seemed that all potential revenue streams had been exhausted, and every day more and more creditors were beating a path to the doors of Shackleton's office. They were no doubt enraged to find a taxi waiting outside all day for his personal use, the meter still running! Among those who were particularly disgruntled were Brocklehurst and Adams, who were yet to receive their wages in full. Adams recalled long afterwards, 'We didn't see a lot of each other, because the whole party, I think they thought they were a bit badly treated on the money side of the business,' and 'I kept calling his attention to the fact that he did owe something to those chaps who had been with him, and that his promises should be kept . . . Of course money meant nothing to him, you see; he didn't know the meaning of the word money except spending it.' Shackleton failed to realize that many of his men had been relying on these promised wages to feed their families.

It seemed that Shackleton was out of ideas, and disgrace beckoned. But then the famed luck of the Irish came to the fore, as it did so often throughout his life when all seemed lost.

When attending the Cowes Week regatta as guests of the

Castle Line, Shackleton became engaged in conversation with Sir Henry Lucy, a respected and well-connected journalist. The two struck up quite a rapport, with Sir Henry fascinated by Shackleton's tales from the Antarctic. Shackleton, the reputable hero, didn't let on about his financial woes, but Emily, speaking to Lady Lucy, did mention the expedition's enormous expenses. When Sir Henry later learned of this, he rode to Shackleton's rescue, writing a story in the *Daily Express* to highlight the fact that one of Britain's greatest heroes was in dire financial straits, due in part to the government failing to back his expedition. An election was fast approaching, and the then Prime Minister, Herbert Asquith, could not risk any public outcry. Bearing this in mind, soon after it was announced that the government would provide Shackleton with a grant of £20,000 (£1.2 million today) in recognition of his great accomplishments for the empire.

Although Shackleton had nothing to do with raising this huge sum of money, and owed it largely to Emily for her conversation with Lady Lucy, he greeted the news as though his business acumen had been proved. 'Just think, your Boy getting £20,000 from the country,' he wrote to Emily. 'What oh!!' This at least allowed Shackleton to finally settle his £1,000 debt to Beardmore, who nonetheless remained in unforgiving mood, as well as most of his other long-standing debts.

During this time, his relationship with Emily became strained. To escape the city, he had moved the family to Norfolk, but had then immediately abandoned them by embarking on his worldwide lecture tour. This was on top of him having already been away for two years. In Norfolk, away from family and friends, Emily understandably felt isolated.

On the rare occasions when Shackleton was at home, it

seems he was unhappy, no doubt troubled by the financial pressures and still dreaming of conquering the pole. Moody and short-tempered, Emily and the children learned to stay out of his way. Yet the moment Shackleton knew he had upset Emily, he would feel dreadful, and do all he could to win her around with his impish sense of humour.

By the turn of the new year, Shackleton was, however, in good spirits, with the wheels in motion for yet another get-rich-quick scheme, which he again assured Emily could lead to a fortune. In January 1910, while in Budapest to address the Hungarian Geographical Society, Shackleton had learned of the Nagybánya mines, where rich seams of gold had been discovered but left unmined. For Shackleton this was too good to be true. If he could persuade investors to back him, they could extract the gold and make a killing. Or so he assumed.

Soon after, at a chance meeting with Mawson, the geologist from *Nimrod*, Shackleton told him about the mines and their potential. Mawson was in fact due to meet with Scott to discuss joining him on his 1911 expedition to the Antarctic, but he was struck by Shackleton's idea.

The two men thrashed out a plan; they would raise the money to buy the mine by floating a newly formed company on the stock exchange. It was also agreed that from this money Shackleton would finally pay Mawson his *Nimrod* wages of £400, and that they would return to the Antarctic at the end of 1911 and explore the uncharted areas of Cape Adare, rather than attempt to reach the pole.

None of this materialized. No company was ever floated on the stock exchange, and when Shackleton tried to persuade his high-society contacts to form a syndicate, he was met with a rude awakening. Lord Northcliffe, the owner of the *Daily Mail*, dismissed the idea out of hand, declaring, 'This is an insult to my intelligence,' and 'I advise you to stick to things you understand.' Neither was there to be a return south in 1911.

When Shackleton presented the proposed expedition to the Royal Geographical Society, without informing Mawson, the two men fell out, and Mawson decided to go it alone. Shackleton had to watch from afar as Mawson set off with Shackleton's closest friend from the *Nimrod*, Frank Wild. This was not, however, too much of a blow, as exploring the uncharted Cape Adare didn't offer the sort of glory, and high stakes, that truly drove Shackleton. What he really wanted, despite having told Emily he would never return, was another crack at reaching the pole.

Knowing that Scott's plans to return to the Antarctic were already well underway, Shackleton tested the water to see if he might be able to mount a rival expedition. To no one's surprise but his own, the RGS offered little support, instead putting their full backing behind Scott. Wilson, his one-time good friend from the *Discovery*, who was also set to return with Scott, attacked his plans too.

Unable to forgive Shackleton for overwintering in McMurdo Sound, despite his promise to the contrary, Wilson was furious when Shackleton mentioned he was looking to make a return. Their friendship already at breaking point, he forced Shackleton to drop his plans until after Scott had his chance to make history. Shackleton reluctantly wrote to Scott soon after and conceded defeat: 'I wish you every success in your endeavour to penetrate the ice and to land on King Edward VII Land and to attain a high latitude from that base.' For now, at least, Shackleton would remain home.

To his credit, despite Scott having tried to lay claim to McMurdo Sound and force Shackleton into a promise not to overwinter there, Shackleton raised no objections to Scott following his path to the pole via the Beardmore Glacier.

Like Shackleton, Scott decided to take ponies, although he ensured they were white, knowing that the majority of Shackleton's had been dark and had died. This was not, of course, due to their colour, but Scott didn't want to take any chances. In addition, Scott took with him thirty-one dogs, believing that if properly trained, they would complement the ponies, especially in hazardous areas where the ponies' weight was against them. In Scott's eyes, Shackleton's shortfall in not quite reaching the pole could be made up with more animal power and more provisions . . . simple mathematics.

In July 1910, Shackleton attended the ceremony at London's Waterloo Station to cheer Scott on his way. Nearby on the platform were Keltie and fellow members of the Royal Geographical Society, who heartily raised three cheers for their great hope, having failed to do the same for Shackleton. It must have hurt Shackleton to know that not only was he considered unfavourably by the RGS, in spite of his achievements, but also that Scott might soon beat his record and consign him to the history books. Nevertheless, as the train departed, Shackleton wished his rival well, not knowing it was the last they would see of each other.

26

Towards the end of 1910, any remaining thought of Shackleton joining the race for the pole was swiftly stopped in its tracks. Emily announced she was pregnant, at which she pleaded to move back to London, where she and the children would be far more at ease.

So, with another mouth to feed, and a more expensive address in Putney Heath to maintain, more pressures were loaded on to Shackleton's already troubled mind and bank account. Emily was now forty-three, and her pregnancy was difficult, so it was inadvisable to leave her alone for any length of time, but Shackleton often had no choice. There were bills to pay and the only way to pay them was by travelling far and wide on his lecture tours.

The pleasure of being acclaimed as a hero and the excitement of new opportunities had worn off. With nothing to occupy his mind, he was now drinking far too much, most notably at the prestigious Marlborough Club, smoking heavily, and piling on the pounds, all of which badly affected his health. On one occasion, he had to be sent home from a business meeting after sweating profusely and complaining of chest pains. While he later put it down to 'rheumatism', Mawson, who was with him at the time, thought it 'must have been angina'.

On the brief occasions when Shackleton was home, he continued to fall into dark moods, which led to more arguments with the increasingly isolated Emily. 'I think I am solely to be blamed,' he admitted, while failing to do much about it. If anything, his mood only worsened, especially with the news now coming from the south.

Already struggling to come to terms with Scott going south, Shackleton had learned that Amundsen was also now making a run for it, having abandoned his attempt to reach the North Pole following Peary's success there. Japan's Lieutenant Nobu Shirase was also planning on trying to reach the South Pole, while the German Wilhelm Filchner was considering the seemingly impossible task of being the first to cross the entire Antarctic continent.

It seemed only a matter of time now before someone beat Shackleton's record. From there, his fame would fade further and further and the lecture tours would be all but over. 'If things [lecturing] do not go better soon I will think very seriously of chucking the whole thing in,' he told Emily. Christy, his lecture agent, also noticed that the sparkle had gone lately and that he 'could be a little hasty'.

Shackleton might have remained desperate to return south, but there was nothing he could do about it. 'The missus is going to have another baby in July and I must not talk of going away,' he wrote, before adding, 'I long for the unbeaten trail again.' Unsurprisingly, when his son Edward Arthur Alexander was born, this did not lead to a renewal of domestic bliss.

It seemed that Shackleton could not sit still for a moment; he was constantly on the move, looking for the next adventure. Being at home only made him think of all he was missing out on. 'I almost wish I had not gone south,' he once said, 'but stayed at home and lived a quiet life.' But now that he had had a taste of adventure, there was no going back.

Emily later said that these were the 'least happy years' of Shackleton's life, adding, 'They certainly were of mine.' She had put up with her husband being constantly absent, his friendships with other women, as well as having to count the pennies when he was away, but now he was home it was the worst of all. Emily was once quoted as saying that she did not believe in reading her children stories that ended with, 'and they lived happily ever after'. She inferred that to do so might encourage a girl to think that marriage was the only option. She then added the opinion, 'How wrong that is.'

Indeed, when helping to choose verses of poetry to head each chapter in Mill's later biography of Shackleton, for this period of his life she quoted from Robert Service's 'Lure of the Little Voices':

> Yes, they're wanting me, they're haunting me, the awful
> lonely places;
> They're whining and they're whimpering as if each had
> a soul;
> They're calling from the wilderness, the vast and
> god-like spaces,
> The stark and sullen solitudes that sentinel the Pole.

Again, Emily was forced to shoulder much of the burden of raising three children, and often paid the bills with her allowance, while her husband was out of the house, apparently searching for other opportunities. 'My wife and 3 children are well,' he wrote to a friend at this time. 'I see little of them though.'

Nothing seemed to cure his itch, certainly not the offer of returning to politics in 1912. Once more, he looked to another woman to soothe and excite him. Rosalind Chetwynd ticked most of Shackleton's boxes. She was ten years younger than him, came from money, being the daughter of a wealthy American lawyer, and she became even wealthier when she married a baronet. Now divorced, she lived in Park Lane, 'surviving' on £5,000 a year from her parents, almost

£300,000 today, while the rent on her apartment was paid for by an admiring mining magnate. Just as he had once pulled out all stops to attract Emily, and then Elspeth Beardmore, Shackleton was now determined to turn the head of Rosalind Chetwynd.

But even the attention of a young, wealthy American lady could not satisfy Shackleton's true craving. 'I wish I could get another expedition,' he wrote in 1912. However, any hope Shackleton might once have had of being the first to reach the pole was shattered on 7 March 1912. On his arrival in Hobart, Amundsen broke the momentous news that he had reached the pole itself on 14 December 1911.

Opting to set up camp at the Bay of Whales, 100 miles closer to the pole than if he had wintered at McMurdo Sound, Amundsen tried a new route to the pole, heading for the uncharted Axel Heiberg Glacier in the hope it offered a shorter and less hazardous journey. For haulage he used fifty-two trained dogs (no ponies) and skis. His gamble paid off, and he reached the pole in thirty-three days. His focus, however, had been on just reaching the pole, rather than being weighed down by any scientific commitments.

On hearing of Amundsen's success, Shackleton noted, 'The dogs will keep up the rapid pace of ski runners and this is naturally faster than the slow-plodding foot movements of the ponies.' If he should ever return to the Antarctic, he swore that this time he would use skis and dogs.

Although disappointed to miss out on the chance to be the first to reach the pole, Shackleton was never one to harbour a grudge. He had come to admire Amundsen and sent him a cable offering his 'heartiest congratulations', while also telling a Norwegian newspaper that Amundsen was 'perhaps the greatest Polar explorer of today'.

To no one's surprise, the Royal Geographical Society

reacted coolly to the news. Their man Scott had been beaten to the pole, and there was still no news of him. Focusing on Amundsen not undertaking any scientific studies on his expedition, the likes of Keltie and Lord Curzon, the new RGS President, rubbished his achievement, calling it a 'dirty trick', and refused to meet with him. Markham also hit out at Amundsen's use of dogs and claimed that Scott's achievement in reaching the pole, via man-hauling, was the real test of a man and 'the true British Way'. This was of course totally untrue, as Scott had also used ponies and dogs. Nevertheless, such was Markham's disgust that when the RGS later finally agreed to host a dinner for Amundsen, he resigned from the RGS Council.

At that same dinner, with Shackleton present to congratulate his rival, Lord Curzon could not resist insulting Amundsen. Rather than asking for three cheers for the explorer, he instead asked for 'three cheers for the dogs'. British newspapers were also unable to accept that a foreigner had beaten Scott to the pole, leading to Amundsen later writing that 'by and large the British are a race of very bad losers'.

Just a few weeks later news finally came of Scott. The *Terra Nova* had arrived at the New Zealand port of Akaroa and the crew claimed that Scott was making good progress, having last been seen (by the last of his support teams before they turned back) in January at 87 degrees 34' S, just 146 geographical miles away from the pole. However, the *Terra Nova* was not set to return to the Antarctic until the summer, meaning that any news of Scott's progress would have to wait for another year. At least, until then, Shackleton could still trade on the title 'Britain's greatest Polar explorer'.

Despite his disappointment, Shackleton soon realized that there was a far more difficult record in the Antarctic yet to be broken: 'The discovery of the South Pole will not be the end of Antarctic exploration,' he wrote. 'The next work of importance to be done in the Antarctic is the determination of the whole coastline of the Antarctic continent and then a transcontinental journey from sea to sea crossing the Pole.' This would entail crossing 1,800 miles, much of which was uncharted territory, in what he described as 'the last great polar journey that can be made'.

At this very moment Wilhelm Filchner was attempting to do exactly the same. So, while Shackleton worked out the logistics of such an expedition, showing the first glimmers of excitement in months, he could not make any concrete plans until he learned whether Filchner had been successful.

The news he craved arrived in January 1913. Filchner, despite making enormous progress, had failed to cross the continent. The record was still up for grabs. Shackleton now told reporters, a glint back in his eye, 'Perhaps I will try again to go south.'

PART FOUR

'Footsteps of courage into stirrups of patience'

– Ernest Shackleton

27

Shackleton was lecturing in America on 10 February 1913 when news of Scott finally broke. The last everyone had heard was that on 3 January 1912 Scott had been well en route to the pole, when Tom Crean had been forced to return to base with Lieutenant Teddy Evans, who was close to death with scurvy. At the time, Scott and his men were just 146 geographical miles from their target, and it was therefore believed that they would reach their destination with ease, soon matching Amundsen's achievement.

However, upon docking at port, the *Terra Nova* carried awful news. Scott and his party of Wilson, Bower and Bates had indeed reached the pole but all had perished on the return journey, ridden with exposure, starvation, frostbite and hampered by poor surfaces and appalling weather. A search party had found their tent, where Scott's last diary entry, dated 29 March 1912, read:

> Every day we have been ready to start for our depot 11 miles away, but outside the door of the tent it remains a scene of whirling drift. I do not think we can hope for any better things now. We shall stick it out to the end, but we are getting weaker, of course, and the end cannot be far. It seems a pity but I do not think I can write more. R. Scott. Last entry. For God's sake look after our people.

The news rocketed around the world. Britain, already at a low ebb following the sinking of the *Titanic* just a year earlier, with the loss of 1,500 lives, as well as the news of Amundsen becoming the first to reach the pole, had now lost one of its most esteemed heroes. It almost felt as if the empire itself were coming to an end. For weeks, stories of Scott dominated the front pages. Cast as a tragic hero, beyond reproach, the nation entered a period of intense mourning, the likes of which had rarely been seen. 'Nothing in our time,' the *Manchester Guardian* declared, 'had touched the whole nation so instantly and so deeply as the loss of these men.'

Four days after the news broke, a memorial service was held in his honour at St Paul's Cathedral, with the likes of Markham, Darwin and Keltie, and the Archbishop of Canterbury in attendance, as well as the King and the elite of British society. Such was the clamour for people to pay their respects, more than 10,000 people gathered outside.

Shackleton was away in America so was unable to attend and it is not recorded how he privately reacted to his rival's death. No doubt he had conflicted feelings over the matter. Scott had given him his chance, but he had also tried to scupper his return south. Perhaps there was a grim sense of competitive satisfaction that the man who had once called him an 'invalid' for struggling in the bitter conditions had himself succumbed to them. This is of course pure speculation, but any number of these thoughts might well have filled Shackleton's mind as he read the eulogies in the American newspapers.

If anything, Scott's death should have been a warning to any intrepid explorer that even the fittest and most prepared could end up dead on the ice. Shackleton had nearly died twice and was now older, and seemingly facing more health problems. Nonetheless, he was not deterred from his mission to return to the Antarctic and cross the continent. In fact, he seemed invigorated. If he should meet the same fate as Scott, then so be it.

The logistics – and the cost – were huge, but after analysing the successes and failures of Amundsen and Filchner, Shackleton hatched a plan he thought might work.

The plan would be to set sail in the summer of 1914 with two ships, each sailing to the opposite sides of the continent. Shackleton's ship would set out from South America and set up camp at Vahsel Bay, via the Weddell Sea, while the other ship would sail from New Zealand and base themselves, via the Ross Sea, at McMurdo Sound. From Vahsel Bay, Shackleton and his party would begin the journey across the ice, carrying enough provisions to last them until the Beardmore Glacier. From there on, the other party would have laid provision depots to sustain them for the rest of the 400-mile journey to McMurdo Sound. Shackleton had taken note of Amundsen's triumph and now decided on 120 dogs and skis as his main mode of transport. Although like Scott, he was also keen to explore using mechanical sledges, whose engines would hopefully help propel them across the ice in record time.

Exactly fifty-eight years later, my wife Ginny's plan to complete the first surface circumpolar journey around Earth would include completing this same geographic feat. However, whereas Shackleton planned to use two ships, to take two expedition teams to opposite coastlines and to meet up somewhere en route, Ginny's plan was for a single ship to drop off a three-man team with three snow scooters, helping to tow two sledges each. We completed this journey in sixty-seven days, becoming the first single team ever to cross Antarctica.

In principle, Shackleton's plan sounded ambitious but exciting. Early estimates, however, suggested it could cost as much as £100,000 (£4.25 million today). If Shackleton had struggled previously to fund the cost of one ship, its crew and provisions, how could he possibly find the money to fund two?

When he had tried to raise the money for the *Nimrod* he had done so without a track record of leading an expedition, and with few wealthy contacts. His semi-success with his *Nimrod* team had not only given him credibility, but now he was feted in high society. Moreover, he was now Sir Ernest Shackleton, a title which immediately granted an element of respectability. In the circumstances, while it seemed that raising £100,000 was a tall order, it was not beyond the realms of possibility. As he said himself, 'All polar explorers are optimists with vivid imaginations.'

However, Scott's tragic death had now tamed the public fervour for yet more polar adventures. Winston Churchill, then First Lord of the Admiralty, was particularly critical, admonishing this pursuit, and writing, 'Enough life and money has been spent on this sterile quest.' The government eventually agreed to provide Shackleton with a £10,000 grant, only half the sum given to Scott, but it was on the proviso that Shackleton first find the remaining £90,000.

Looking for other donors, Shackleton was well aware that it was very unlikely there would be any significant contribution forthcoming from the Royal Geographical Society. In a letter to William Speirs Bruce, the Scottish polar explorer, he wrote, 'I cannot look, nor am I going to try, for assistance from the Royal Geographical Society. You know as well as I do that they are hide-bound and narrow and that neither you nor I happen to be particular pets of theirs.' Despite this, Markham still could not help but stick his oar in, calling Shackleton's plans 'useless and expensive', and 'designed solely for self-advertisement'. He also doubted that Shackleton was the man to undertake such an ambitious expedition.

In fairness to the RGS, Scott had only just been put to rest and Shackleton's plans did seem outlandish. Even Mill, Shackleton's great and loyal friend, refused to give his approval for his transcontinental plan. In his eyes, the whole frozen desert was bigger than Europe, the USA and Mexico combined, yet Shackleton was hoping to cross it in just a hundred days, aiming to walk 16 miles a day, often through large chunks of uncharted territory. The Weddell Sea was also known to have unpredictable pack ice all year round, with Mill warning that, 'No two voyagers have found similar conditions.' Indeed, Filchner's expedition had to be abandoned after his ship, the *Deutschland*, was trapped in pack ice and carried away entrapped for over nine months. Shackleton also couldn't count on gaining a good anchorage at Vahsel Bay sufficiently early in the short southern summer to have enough time to complete the journey to the Ross Sea side before winter set in, even assisted by motor sledges and dogs. In short, it was too hazardous and over-ambitious.

I well recall the similar criticism we received when we set out to circumnavigate Earth via both poles in the late seventies. Sir Miles Clifford, one of Britain's most respected polar veterans and gurus, admonished me for my plans at the time, telling me, 'You are saying, Fiennes, that your group will, in the course of a single journey, complete the great journeys of Scott, Amundsen, Nansen, Peary and Franklin and many others? You must understand that this sounds a touch presumptuous, if not indeed, far-fetched.' Of course, this was entirely the idea, but like Shackleton, to make such a dream a reality, we often had to survive on nothing more than an unshakeable belief that we could achieve the impossible. And the more people told us it could not be done, the more we were determined to prove them wrong, just as Shackleton now hoped to do.

Despite calling Shackleton's plans 'audacious in the extreme', in an attempt to save some face the Royal Geographical Society did eventually provide a grant for £1,000. It certainly didn't want to be in a position where, if Shackleton should again surprise them, they were seen to have offered no support at all.

Royal support had been a game changer for the *Nimrod* quest, and to Shackleton's delight King George V now backed his Imperial Trans-Antarctic Expedition (ITAE). With this, Shackleton felt he could unveil his plans to the public, hoping to garner further support, and money.

On 29 December 1913, he revealed his intentions in *The Times*, hoping to spark a rush of willing donors to his door. *The Times* certainly beat the drum on his behalf, exclaiming that following the 1912 disasters of *Titanic* and Scott, Shackleton was aiming to 're-establish the prestige of Great Britain'. Yet once more, Frank Shackleton's antics threatened to put the whole expedition in peril.

Never seeming to stray far from trouble, Frank Shackleton was put on trial for fraud, accused of stealing £1,000 from Mary Browne, an elderly spinster. After Frank had been found guilty, and sentenced to fifteen months in prison, Shackleton was lucky to escape any guilt by association. The co-accused was Thomas Garlick, who had been Shackleton's partner in his attempt to bring home Russian soldiers during the Russo-Japanese War. In any event, the well-publicized trial, and Frank's imprisonment, didn't help his brother in his fundraising. To avoid becoming embroiled in the negative publicity, Shackleton had no choice but to settle out of court with Mary Browne. He said of all of this, 'My brother has been more of a fool than anything else, it has struck my business as you can imagine.'

Meanwhile, crossing the Antarctic continent was now

seen as the last great polar prize, and the threat of rival expeditions now reared their head. The most feared of these rivals was the Austrian Dr Felix König, who had been on Filchner's crew. König not only announced that he would also be making his attempt to cross the continent in 1914 but, to Shackleton's dismay, his plans were already far more advanced than his own. Unlike Shackleton, König had already acquired significant financial support, thus putting further pressure on Shackleton to hurry along his own preparations.

I remember during many expeditions being in a supposed race against other competitors and feeling I must do anything to beat them. However, I was very fortunate to have Prince Charles in my corner to hand me some sage advice. 'No racing!' he told me, making it clear that we should be well prepared and look to achieve our goal to the best of our ability, no matter whatever anyone else was doing. This advice was to serve me well, and while of course I still felt the competitive instinct to be the first, I also ensured that I did not cut any corners, as would prove to be so costly for Shackleton.

With König looking to set sail around the same time as Shackleton, the two soon clashed over proprietorial rights. Believing that he had established rights to the Weddell Sea's southern coastline by virtue of being the first to explore it with Filchner, König tried to prevent Shackleton from using it. He also told him that he would be using Vahsel Bay as his winter camp and Shackleton should look elsewhere.

In his dispute with Scott, Shackleton had recognized that he owed some sort of debt to his former leader. Now, older and more experienced, Shackleton would not give an inch to König. 'I cannot alter plans I have long since formulated,' he told him. To the Royal Geographical Society he wrote, 'I have as much right to use [Vahsel Bay] as Dr König.' But if Shackleton did not raise the funds soon, he would have no choice but to sit back and seethe as König set sail.

By now donations were at least pouring in, thanks in part to Shackleton's announcement in *The Times*. Frank Dudley Docker, a wealthy industrialist from Birmingham, offered up £10,000, and Sir James Caird added the same. The seventy-seven-year-old Caird, who had made his fortune as a textile manufacturer in Dundee, would soon add another £24,000, with no strings attached. Shackleton had always dreamed of finding a wealthy donor who could provide a substantial chunk of the funds in one swoop, as Longstaff once had for Scott. Finally, he had found his man.

Unbelievably, the rest of the money soon followed. Elizabeth Dawson-Lambton could always be relied upon to swell the coffers with a gift, while Shackleton's persuasive charm helped to prompt yet another wealthy woman to provide further contributions. Janet Stancomb-Wills was in her early sixties, had inherited half of the estate of her millionaire father and was no stranger to making charitable contributions. Disarmed by his charm and tales of adventure, she wrote to Shackleton, 'Into my life you flashed, like a meteor out of the dark.'

The pair soon became firm friends. 'I have hammered through life, made but few friends and it is good to know you,' he told her, in the now-familiar refrain. Stancomb-Wills was to make another notable donation to the expedition fund. Now, with almost all the money he needed, Shackleton could turn his attention to making the biggest purchase of the expedition: two ships.

The first purchase was the *Polaris*, a three-masted ship built in a Norwegian shipyard for the apparently wealthy owner of a whaling fleet, who was now unable to pay for it.

Shackleton swooped in and agreed to pay cost price, some £11,600. He even struck a deal to pay in instalments, freeing up more of his funds, and changed the name of the ship to *Endurance*, in a nod to his family motto: 'By endurance we conquer.' This would be the ship that Shackleton himself would travel on, hopefully through the Weddell Sea to Vahsel Bay.

The second ship, the *Aurora*, thankfully cost far less. At over forty years old, she had spent much of her life as a whaling ship, before being used by Mawson for his Australasian Antarctic Expedition. Mawson now had no more use for the ship and desperately required funds to meet costs. He therefore agreed to sell it to Shackleton for the knock-down price of £3,200. The *Aurora* would thus be used by the party who would travel through the Ross Sea to McMurdo Sound.

Next on the agenda was the crew. This was a huge task, as Shackleton now had to find suitable men for two ships, both on very different journeys. As before, he initially fell back on men who had been tried and tested in such conditions. Frank Wild, to whom Shackleton had become so close during the *Discovery* and *Nimrod* expeditions, was certainly one of those. Now back from his expedition with Mawson, he was quickly assigned as Shackleton's deputy, and put to work in the Regent Street office. There, he sifted through over 5,000 applications to join the crew, dividing them into three categories for Shackleton's consideration: 'Mad', 'Hopeless' and 'Possible'.

It is not known on which pile Wild put the application of his younger brother, Ernest, but nevertheless Ernest was selected to join the Ross Sea party and put in charge of training the dogs. Those who were definitely not placed in the 'Possible' file included all members of the opposite sex.

After receiving an application from 'three sporty girls'

who did not see 'why men should have all the glory and women none', Shackleton responded that there were 'no vacancies for the opposite sex on the Expedition'. This outdated attitude flies in the face of my own expeditions, where I only wanted the best people for the job, regardless of whether they were men or women. Indeed, my wife, Ginny, was the first woman to win a Polar Medal for her research work on our expeditions.

Many individuals Shackleton knew well from *Discovery* and *Nimrod* days swiftly came on board. Tom Crean was appointed as second officer on the *Endurance*, while Aeneas Mackintosh was selected to take charge of the Ross Sea landing party, who would place the vital provision depots on the Barrier for Shackleton to pick up while crossing. Ernest Joyce, who had sailed with Shackleton on both the *Nimrod* and the *Discovery*, was added to the team, as was George Marston, who would again serve as expedition artist.

Day by day the crew slowly came together, leaving Shackleton to concentrate on one of the most important jobs: finding captains for the *Aurora* and the *Endurance*. Following the debacle with *Nimrod*'s Captain England, and with Captain Evans not faring much better in some eyes, Shackleton had his work cut out to find one suitable captain, let alone two.

His disagreement on the bridge with Captain England had become headline news, with Captain England's reputation, and career, severely harmed as a result. Shackleton had subsequently ensured that England was the only member of the party not to receive a silver Polar Medal on their return. All of this kept away many prospective candidates, especially as there were also whispers that Shackleton could not be relied upon to keep his promise in regard to pay and bonuses. Bearing this in mind, John King Davis, Shackleton's first choice to captain the *Endurance*, turned him down. While this seemed a blow, it was in fact an extraordinary bit of luck.

Without Davis at the helm, a Merchant Navy officer by the name of Frank Worsley now came into the reckoning. Worsley was in London on leave when news of the expedition caught his eye. Shackleton was immediately taken by the forty-two-year-old New Zealander. He certainly had a wealth of experience at sea but, far more than that, he appeared to be the sort of adventurous personality who would suit the unique pressures of sailing in the Antarctic. As a bonus, he was excellent company, almost an equal to Shackleton when it came to the ability to tell an enthralling story. There was no question about it. Worsley was the man to captain the *Endurance*.

Yet while Shackleton was fortunate to find a suitable captain for the *Endurance*, finding the same for the *Aurora* was to prove more difficult. Although Mackintosh would be in command of the ship, an officer was still required to stand in when he was on the ice. With suitable candidates in short supply, Shackleton wrote directly to Winston Churchill, First Lord of the Admiralty, and asked if he might be prepared to release any captains from the Navy.

However, on 28 June 1914, Archduke Franz Ferdinand was assassinated in Sarajevo. Soon after, war was declared between Austro-Hungary and Serbia, with Germany backing the Austro-Hungarians. A major conflict looked unavoidable and, as such, Shackleton's appeal to Churchill was met with a muted response. Captains from the Royal Navy were now off limits, and Shackleton's pool of potential captains greatly diminished.

The shortage of suitable candidates led Shackleton to consider someone whom he had previously rejected. Joseph Stenhouse was a captain with the Merchant Navy, but the twenty-six-year-old was known to suffer from depression. This was not a characteristic necessarily suited to the rigours of the Antarctic, but Stenhouse was persistent, and kept on pushing his case. In the end, Shackleton had little choice. Stenhouse was hired to captain the *Aurora* once Mackintosh took command of the landing party at McMurdo Sound. He was over the moon: 'I am a lucky chap!'

The advent of war had other consequences for the planning of the expedition. Shackleton had purchased 120 dogs from Canada, but just 99 had arrived in London by July 1914. With an average weight of 100 lbs, Wild said they were a mixture of 'wolf and almost any kind of big dog'. To train them, Shackleton had appointed Lieutenant F. Dobbs of the Royal Dublin Fusiliers, but Dobbs was forced to resign, given the worsening situation in Europe. Shackleton turned to two Icelanders, Jón Björnson Johnsson and Sigurjón Isfeld, to fill the position, but both men were newly married, and expectant fathers, and declined.

Sir Daniel Gooch, an experienced breeder of dogs for hunting, yet with no experience on the ice, was Shackleton's next choice. He was, however, only able to travel with Shackleton on the *Endurance* as far as South Georgia. Until then it was down to Wild to watch and learn from him.

As with the *Nimrod*, Shackleton's interview process was a touch quixotic. Dr Alexander Macklin, then a surgeon in a Blackburn hospital, recalled with bemusement the first time they met. 'One question was, "Is your eyesight all right?" I was wearing specs. I said it was. He asked, "Why are you wearing spectacles?" For want of anything better, I said, "Many a wise face would look foolish without spectacles," and he laughed.' Macklin was subsequently hired as the surgeon on the *Endurance*.

Dr James McIlroy recalled his strange interview for the position of second surgeon:

> Shackleton could be a very frightening kind of individual; like Napoleon, he was very stern looking and fixed you with a steely eye. I wasn't asked to sit down. I stood in front of him, facing the light . . . He asked me lots of questions. One was, 'Where have you just come from?' I said the Malay States. He said, 'Well, I suppose you know all about malaria and the other fevers they get out there?' I said, 'Yes, I'm supposed to know about them.' He said, 'Well, of course, that's no damned good to me, we won't get anything like that down in the Antarctic! Have you ever had any experience of the cold?' I had to admit I hadn't, apart from the cold one gets in England. However, he said, 'Well, I'd like you to be examined. You seem to be shaking a lot.' I said, 'I'm a bit nervous in front of you.' As a matter of fact, it was malaria! So I went and saw a physician and he gave me a very good report, and Shackleton was quite content with that.

Dr Reginald James, who applied for the position of physicist on the *Endurance*, remembered:

> I was appointed after an interview of about ten minutes at the outside, probably more nearly five. So far as I remember he asked me if my teeth were good, if I suffered from varicose veins, if I had a good temper, and if I could sing. At this last question I probably looked a bit taken aback, for I remember he added, 'Oh, I don't mean any Caruso stuff; but I suppose you can shout a bit with the boys?' He then asked me if my circulation was good. I said it was except for one finger, which frequently went dead in cold weather. He

asked me if I would seriously mind losing it. I said I would risk that . . . After this he put out his hand and said, 'Very well, I'll take you.'

Rounding off the crew was twenty-nine-year-old Australian photographer Frank Hurley, who had just returned from a year on the ice with Mawson. For his latest money-making scheme, Shackleton had set up a new company known as the Imperial Trans-Antarctic Film Syndicate. Owning the rights to all footage shot on the expedition, he aimed to sell Hurley's film and photographs to the media, and use them for his own lectures. Shackleton was truly a pioneer in this respect, having recognized before many others the value of good publicity and controlling the narrative.

Soon after, Lionel Greenstreet was appointed first officer, while Scotsman Henry 'Chips' McNish came on board as the ship's carpenter. Shackleton was later to say that the Scotsman with a short temper and an unwillingness to hold his tongue was 'the only man I am not certain of'. In addition, ex-naval fishing-trawler hands from Hull, Grimsby and Labrador were to join at the last moment.

Of particular importance to Shackleton was food. It had continually plagued him that if his *Nimrod* team had just had 'another 50 lb of food', they would have reached the pole. Moreover, while on the *Discovery* he had almost died of scurvy, due to lack of vitamin C. This time around, he wanted to be far more prepared when it came to nutrition than any expedition to date.

At the time, the best person to speak to about such matters was Colonel Wilfred Beveridge, a professor of hygiene at the Royal Army Medical College. Beveridge was an army nutrition expert and had a wealth of experience in how to feed armies while on the march. On Beveridge's recommendation,

Shackleton took with him special 'cakes' made of compressed oatmeal, sugar, beef powder and proteins, and 'nut food' consisting of powdered milk, sugar, Marmite, tea and salt. These dehydrated rations were the first of their kind to be taken on a polar expedition, and while they would help to ward off scurvy they were also far lighter than other rations.

The ships, dogs, provisions and crew were coming together, albeit in a typically chaotic fashion. However, Shackleton's relationship with Emily became more and more strained. He would once again be leaving her for at least another twelve months, with little money and three children to raise by herself. In the months leading up to departure he had rarely been home, making matters worse. 'I know that if you were married to a more domesticated man you would have been much happier,' he told her, recognizing his flaws. 'I am just good as an explorer and nothing else.' As always, he clung to the belief that if, this time, he was successful, he would return a changed man: 'I am going to carry through this work and then there will be an end I expect to my wanderings for any length of time in far places.'

Emily had heard this all before. As she said herself, 'He always said each expedition would be his last, can you wonder I believed it?' Yet still she did not try to stop her husband. There would be no taming him, or his dreams. 'When a man's heart is set in that direction,' she said, 'and especially when he is so suited for the work, one has to put one's own feelings aside – though you know how terribly hard it is sometimes.' Years later, Ginny once said something very similar about me, telling *Woman's Weekly*, 'I was brought up in a fairly strict, old-fashioned, stiff-upper-lip kind of way where you don't show emotions. You don't just sit down and weep. You have to get on with things . . . I've never said, "Don't go" and I never will.' But I was luckier than Shackleton. For many years, before she fell in love with cattle farming, she was an integral part of my expeditions.

Miraculously, the expedition was all but ready to depart by August 1914, with the dowager Queen Alexandra once more paying Shackleton a visit on the eve of departure. On the deck of the *Endurance*, the sixty-nine-year-old dowager Queen presented Shackleton with a silk Union Jack, as well as two Bibles, inside one of which she wrote, 'May the Lord help you to do your deeds, guide you through all dangers by land and sea. May you see the works of the Lord and all his wonders in the Deep.'

However, as the *Endurance* set sail for Ramsgate on 1 August, Shackleton was concerned that the expedition might have to be abandoned. On 1 August, Germany had declared war on Russia, and then, on 3 August, as the *Endurance* was moored at Ramsgate, Germany had then declared war on France. All-out war was now inevitable. In response, Britain ordered general mobilization for the armed forces. Four men on board the *Endurance* immediately resigned and went off to war. König's expedition was also affected. His ship was moored in the Austro-Hungarian port of Trieste, which meant he was prevented from travelling south and was instead forced to enlist in the German army. It seemed that Shackleton and the rest of his crew would meet a similar fate.

Shackleton knew that it would be impossible to depart without putting his ship, and crew, at the disposal of the War Office. Sending a cable to the Admiralty, offering the same, he had to come to terms with the fact that the expedition was over, just as it was set to leave Britain.

Informing the men that they were free to enlist, he wondered where he might now end up himself? Wherever it was, it would be far away from the tranquil white landscape of the Antarctic. Indeed, he also wondered whether, if he were to survive the war, he would ever again get the chance to go south. By then, he would surely be too old, and the prospect of once more raising the money seemed daunting. However, as these thoughts whirred around in his head, a cable from the Admiralty was thrust into Shackleton's hand. It was from Churchill himself, and contained just one word: 'Proceed.'

I often wonder if my grandfather Eustace Fiennes, who was Churchill's Personal Assistant at around this time, was involved in this decision.

28

Not waiting around for Churchill to change his mind, both the *Endurance* and the *Aurora* immediately set sail for their respective ports, with the *Aurora* heading to Hobart and the *Endurance* for Buenos Aires. Shackleton, meanwhile, stayed behind to tie up some loose ends before boarding a mailboat to South America to catch up with his ship. If he thought he would be leaving his troubles behind in England, he was, however, to be disappointed.

Arriving in Buenos Aires in mid-October, Shackleton was immediately met by chaos. Despite being a new ship, the *Endurance* was already undergoing substantial repairs in port, having sprung a leak on the journey across. Indeed, she was lucky to have even arrived. Her coal bunkers had just half of the *Discovery*'s capacity, and two days away from port the ship ran out of coal. In order to reach the closest port of Montevideo, all spare wood, including some of the masts, had to be burnt as fuel. Time was of the essence if they were to get to the Antarctic before the ice blocked their path, but the Argentinian authorities seemed intent on taking their time fixing the leak. Taking matters from bad to worse, Shackleton also learned that some of the crew were also in hot water.

At a stopover in Madeira, some had turned to drink and wrecked a local café during a brawl. Four of the crew were thrown in jail, with one flogged, and four more, including the cook, were sacked by Worsley. Yet with time on their hands in Buenos Aires, many of the crew were now drinking again, apparently ignoring Worsley's protests. This did not bode well. Shackleton would have to stamp his authority on proceedings, to stop anything similar happening again.

Firstly, in need of a cook, he swiftly appointed Charles Green, who by good fortune was in Buenos Aires at the time. Perce Blackborow, a young stowaway found hiding in a locker when the *Endurance* eventually left Buenos Aires, was signed on as the cook's assistant, but only after putting a grin on Shackleton's face. While Shackleton was in the process of tearing strips from him for daring to stow away, the Boss made the following mock threat, as Wild recalled:

> 'Do you know that on these expeditions we often get very hungry, and if there is a stowaway available he is the first to be eaten?' Shackleton was . . . fairly heavily built, and the boy looked him over and said, 'They'd get a lot more meat off you, sir!' The Boss turned away to hide a grin and told me to turn the lad over to bo'sun, but added, 'Introduce him to the cook first.' Blackborrow turned out to be a good sailor . . . and was duly signed on.

Aiming to get the *Endurance* shipshape as soon as possible, Shackleton managed to speed the repairs along with his mix of charm and eager purpose. 'I think he could persuade anyone to do almost anything if only he could talk to them,' the physicist Reginald James recalled.

The *Endurance*'s coal bunker capacity was raised to 160 tons, but Shackleton now found that the cost of fuel in Buenos Aires was far more than he had budgeted for. At the current rate, he couldn't even afford enough coal to get her to her next port, South Georgia. Humiliatingly, Shackleton was forced to borrow $25 from James Wordie, the expedition geologist. 'It does not amount to very much,' Wordie wrote in his diary, 'but will get him out of a hole without raising trouble in London.'

Perhaps most concerning of all, amidst this carnival of chaos, was Shackleton's health. His previous two expeditions had pushed him to the brink of death, and any doctor who was able to examine him expressed concerns about his heart and breathing difficulties. Now, he fell ill with what he called 'suppressed influenza', blaming his hectic work schedule for his ailments. It is impossible to say what was truly troubling Shackleton, as once more he refused to allow the expedition's doctors, James McIlroy and Alexander Macklin, to examine him.

As Shackleton tried to keep the expedition on track, he was coming in for fierce criticism in the British press. *Bystander*, a society journal, summed up the general feeling by writing that it was a 'lost opportunity' when 'Sir Ernest Shackleton decided to go not to the front but to the South Pole.' The criticism stung Shackleton. He had done everything in his power to put himself, his men and his ship at the War Office's disposal and had been told to continue with his expedition. Indeed, before departing for Buenos Aires, he had met the War Minister, Lord Kitchener, who had also given his approval. Moreover, even if he had wanted to fight, at forty years of age, he was too old for conscription. Besides, many whom he spoke to were also certain that the conflict would be all over by Christmas. He certainly didn't feel any guilt, but he hated the thought of the media not being on his side. Still, he had more than enough to take his mind off a few unkind words.

Across the other side of the world, the *Aurora* wasn't faring much better than the *Endurance*. Shackleton had promised Mackintosh that £1,000 would be waiting for him in Hobart, with which he could purchase provisions for the depot-laying operation and keep the ship anchored in McMurdo Sound for up to two years. However, this was only half of what was

24. The *Endurance* crew.

(a.) Shackleton was immediately taken by the forty-two-year-old New Zealander Captain Frank Worsley. He had a wealth of skill, but more than that, he was the sort of adventurous personality who would suit the unique pressures of sailing the Antarctic. As a bonus, he was excellent company, almost an equal to Shackleton when it came to telling a good story – just the man to captain the *Endurance*.

(b.) Shackleton knew Tom Crean well from *Discovery*. A fellow Irishman and former Royal Navy man, with an ability to endure inconceivably unpleasant conditions with good humour, he was appointed as second officer (and cook) on the *Endurance*. Crean trusted Shackleton's instincts and never questioned his leadership. He was rewarded for this loyalty by being selected to cross South Georgia with Shackleton and Worsley.

(c.) Frank Wild was appointed as Shackleton's deputy, after the two men became close on the *Nimrod* and *Discovery*. The ever-faithful Wild was able to get on with all the men, whatever their station, without losing his authority. The men respected and trusted him.

(d.) Rounding off the crew was the twenty-nine-year-old Australian photographer Frank Hurley, having just returned from a year on the ice with Mawson. Hurley was solely responsible for the expedition photographs. He was an artist that 'would go anywhere or do anything to get a picture'.

25. (a.) The last attempt to break out of the ice, 14–15 February 1915. The battle against the ice was incessant. (b.) Charlie Burton cutting up an ice floe to use as a raft to cross an open canal during the authors Trans Globe Expedition in 1982.

26. In 1982, the author's Trans Globe Expedition ship beset in the ice floes.

27. The author and Oliver Shepard on moving ice floes waiting for freeze-up in 1977.

28. Frank Wild observes the sinking of *Endurance*.

29. Shackleton oversees an attempt to manhaul lifeboats as sledges to escape to the nearest landfall following the sinking of *Discovery*.

30. After abandoning *Endurance*, the men were unable to reach dry land, so they found a solid floe on which to set up camp, christened Ocean Camp. The men were assigned to five tents while Shackleton faced some grim realities – risk of drifting out to sea, or the floe collapsing around them.

31. (a. and b.) Saying his goodbyes to the men who would be left behind on Elephant Island, Shackleton felt the weight of responsibility. They were all relying on him, and he was determined to get every single one of them home. 'Skipper, if anything happens to me while those fellows are waiting for me, I shall feel like a murderer,' he told Worsley.

32. The men left behind on Elephant Island made a 'hut' out of the two remaining lifeboats, sheltering twenty-two men for four months.

33. (a. and b.) Our navigation methods during our crossing of Antarctica in 1979–80 were exactly the same as Shackleton's. There were no polar-orbiting satellites, no satnav, no GPS (nor smartphones). To locate our position we used, as did Worsley, a theodolite or sextant.

34. Crean, Shackleton and Worsley with Captain Thom (*second from right*) in Husvik, photographed a day after their return to civilization. Bathed, shaved and freshly clad, Shackleton's immediate concern was for the men on Elephant Island, who he was determined to rescue with all possible haste.

35. Shackleton's party on the *Endurance* heading for the Weddell Sea, while the other party on the *Aurora*, under the leadership of Aeneas Mackintosh, started from the Ross Sea and set out to lay depots to meet the Trans-Arctic team.

36. Shackleton leading the *Aurora* relief party, searching for the Ross Sea party. While they found six men on their way back from Cape Evans, three others, Mackintosh, Spencer-Smith and Hayward, had died. Shackleton made a search for the bodies, but without success, and *Aurora* sailed back to New Zealand. Shackleton could no longer state 'I never lost a man'.

37. Ernest Shackleton as he was just before his *Quest* voyage, aged forty-seven.

38. On 17 September 1921, *Quest* set sail from St Katharine Dock, stearning under Tower Bridge, en route for the south on the Shackleton-Rowett Expedition.

39. On 2 September 1979, the *Benjamin Bowring* sets out on her three-year voyage to complete the ocean phases of the author's circumpolar surface journey.

40. In May 1922, Wild, now in charge of *Quest*, brought the expedition back to South Georgia to visit Shackleton's grave.

41. To commemorate Shackleton's death on 5 January 1922, his comrades erected a cairn surmounted by a cross on the slopes of Duse Fell in South Georgia.

required, and Shackleton had told Mackintosh, 'You will economise in every way.' For Shackleton, this meant bluffing and blagging more sponsorship and freebies, while also utilizing anything that had been left behind in the huts at Cape Royds and Cape Evans from previous expeditions. Things were already on a knife edge, but on his arrival at Hobart, Mackintosh found that not only was there was no £1,000 waiting for him, but expenses were far more than expected.

He was dumbfounded. 'How coal and equipment are to be obtained,' he wrote, 'the very large deficiency of stores made up, and wages and advances to the crew paid is not indicated . . . The ship is held up, valuable time is being lost, and I have very grave fears for the result.'

Mackintosh urgently cabled London. 'Money I must have . . .' he urged. 'There are lives of my men on my hands.' The sum of £700 eventually arrived from London, but Shackleton and Mackintosh once more had Professor David to thank for getting them out of a jam. Having persuaded the Australian government to provide a grant of £500, David also arranged for a mortgage on the *Aurora* to raise more funds. The ship was also opened to the public, who responded with countless gifts, such as items of food, books and crates of whisky. John King Davis, who had previously been asked by Shackleton to captain the *Endurance*, was also a tremendous help, arranging for a sponsored wireless set to be installed on board, among other things. This addition would later prove crucial. By the skin of his teeth, Mackintosh was able to purchase the bare minimum of equipment and supplies, but the disorder was a sign of things to come.

In Buenos Aires, Shackleton had by now made a partial recovery, and with the *Endurance* fixed, and sufficient coal in the bunker, she finally set sail for South Georgia on 26 October. Shackleton let out a sigh of relief. 'All the strain is finished and there now comes the actual work itself,' he wrote '. . . the fight will be good.'

With every mile south, Shackleton felt the thrill of excitement, the prospect of redemption from past failures. Until this great venture was a reality, he had feared that, like so many ex-sportsmen, he would fade into obscurity once his bow was seen to be firing no new arrows. Now, back in command of an expedition, he felt young and purposeful again, coming alive among the men, sharing stories and jokes and looking forward to the chance to make history.

On 5 November 1914, the murky island of South Georgia, 'the gateway to Antarctica', emerged before them in the mist. While Captain James Cook had claimed the mountainous island for Britain in 1775, much of the interior remained largely uncharted, although its harbours were now home to a thriving Norwegian whaling industry. Arriving at the blood-stained port of Grytviken, Norwegian sailors could be seen aboard whaling ships, each carrying enormous whale carcasses, the smell of which immediately tinged the nostrils. This desolate island, the furthest outpost of the British empire, with its small shanty town the only sign of civilization, would serve as the *Endurance*'s home for the next month.

Still in need of coal and provisions, but all out of money, Shackleton shamelessly bluffed over £400 of credit for winter clothing, butter and flour. He achieved this by asking for the bills to be sent to Ernest Perris, his agent in London, aware that Perris did not have sufficient funds at hand.

Shackleton's original plan had been for his five-man party, consisting of himself, Wild, Crean, Marston, Hurley and Macklin, to start the crossing as soon as they arrived at Vahsel Bay, aiming to have completed it by March 1915. In the Antarctic summer, it was possible to travel from early November to February, but owing to the time spent in Buenos

Aires they had missed that window. They would now need to overwinter at Vahsel Bay, with the *Endurance* being purposefully trapped in the ice, just like the *Discovery* had once been.

This information needed to be cabled to Mackintosh in Australia. There was no point in his party commencing its depot-laying operation until Shackleton belatedly began the march, as the weather in the intervening months might bury or sweep the depots away. Yet while Shackleton contended that a cable to Mackintosh was sent, Mackintosh never saw it, and would begin the depot-laying operation as soon as the *Aurora* landed in Cape Evans.

In this muddle of confusion, on 5 December 1914, the *Endurance* left Grytviken and made her way towards Vahsel Bay. Somewhat prophetically, Shackleton chose to write a few lines in his diary from the poem 'The Ship of Fools' by St John Lucas, adding some of his own amendments:

> We were the fools who could not rest in the dull
> earth we left behind.
> But burned with passion for the South.
> And drank strange frenzy from the wind.
> The world where wise men sit at ease,
> Fades from my unregretful eyes
> And blind across uncharted seas
> We stagger on our enterprise.

Jim Mayer, in *Shackleton: A Life in Poetry*, reveals that 'The Ship of Fools was a title first used in 1494 for a poem by Sebastian Brandt that presents an allegory in which fools steer a ship full of more fools towards a fools' paradise.' There were certainly plenty who might have felt the same about Shackleton and his expedition.

Just reaching Vahsel Bay would prove to be a far greater challenge than Shackleton had envisaged. Since James Weddell had first sailed through the area ninety years previously, few ships had followed. Those who had, including William Speirs Bruce's *Scotia* in 1902, and Filchner's *Deutschland*, were almost crushed in the ice.

As soon as the 57th parallel, Shackleton found that groaning and screaming pack ice was already blocking the way. This was much further north than whaler captains had reported for December in previous years. And there was still 1,000 miles of this treacherous labyrinth ahead, much of which was impenetrable, and poised like a hostile rat trap.

In 1972, when Ginny first planned a route for our circumpolar journey, she studied the behaviour of the Southern Ocean's different zones and quickly decided to avoid the Weddell Sea, from which, she noted, 'Soviet sailors had in 1965 measured a berg, split from an ice shelf, which was 87 miles long and 2,700 square miles in size.'

As the *Endurance* struggled to smash her way through to Vahsel Bay, Shackleton wrote, 'This ship is not as strong as the *Nimrod*. I would exchange her for the old *Nimrod* any day now except for comfort.'

Finding a route through the ice was like trying to work your way through a maze. One moment there seemed to be newly open water, emanating dark clouds of frost-smoke, the next the ice would again converge, blocking any way forward. Thankfully, Worsley was in his element. Directing steerage of the *Endurance* with verve, either from the crow's nest, or at the wheel, he sneaked between open leads, colliding with stubborn floes where necessary, always attempting to head in a southerly direction. On those occasions when progress appeared all but impossible and there was real concern they might have to turn back, the ice would often open up and beckon them forward.

Despite zigzagging their way through the ice, reaching

Vahsel Bay was soon beginning to look impossible. Worsley was keen to find an alternative coastal base, and on 15 January 1915, Shackleton sighted a glacial bay. While it looked a good place to set anchor, Shackleton realized it would add another 200 miles to his overland journey across the continent. He decided that this was too much, and told Worsley to continue further south, notwithstanding the treacherous conditions. It was a decision he would come to regret.

On the morning of 19 January 1915, the crew woke to find that the ship was no longer making headway. They were trapped, with no apparent cracks or open water to aim for. Rather than overwintering on land, they were now at the mercy of the ice, with millions of tons pressing against the *Endurance*'s creaking hull.

29

It quickly became apparent that, for the winter at least, there would be no escape from the ice. All efforts were now directed into making the next few months as comfortable as possible for the twenty-eight men on board.

Of primary concern was saving the dwindling coal supplies; just 67 tons remained, enough for thirty-three days of travel. This would get them to Vahsel Bay, just 40 miles away, but having enough to return to South Georgia was another matter altogether. After some agonizing, Shackleton realized that as soon as they were free of the pack, they had no choice but to immediately return to South Georgia to purchase more coal before setting out again.

The *Endurance* was now drifting with the moving pack ice, ever further from its destination, and further into the unknown. 'My chief anxiety is the drift,' Shackleton shared with his diary. 'Where will the vagrant winds and currents carry the ship during the long winter months ahead of us?' But there was little he could do. He just had to pray the drift would be kind and not take them as far away as it had Filchner's *Deutschland*, which had drifted for over 700 miles.

At one point he considered marching to Prince Regent Luitpold Land, the nearest landfall, which was only 20 miles away. However, marching across the hazardous sea ice would be extremely risky, and it would also be impossible for the men to carry all of the equipment and provisions. After some thought, the idea was scrapped. He would just have to be patient.

Rather than burden the men with these worries, or allow

their minds to wander through inactivity, Shackleton fell back on his experience from previous expeditions. First and foremost, he wanted to avoid what had happened to the crew of *Deutschland* in their months trapped in the ice. The ship's captain, Richard Vahsel, had died, and the crew were beset by feuds, with one recalling that the crew had 'the worst interpersonal frictions and dissensions'.

Now that the Antarctic winter had set in, plunging the boat into darkness for the next three months, Shackleton knew from experience that this could adversely affect his crew's mood. He therefore strove to remain as optimistic as possible. Always seeking to lighten the mood, he went from group to group, eager to show no favouritism, leaving laughter in his wake. Worsley said that the man known to them all as 'Boss' was a 'cheery chief, leading his men in a great adventure', while former Royal Marine Thomas Orde-Lees, the expedition storekeeper and ski expert, praised him for bringing the men together and not allowing cliques to form.

Orde-Lees also remembered one particular act of kindness which would not have surprised anyone who had previously travelled with Shackleton. After a long and cold solo trek on the sea ice, Orde-Lees suffered from an extremely painful bout of sciatica, a condition that Shackleton had himself long suffered. Recognizing Orde-Lees's pain, he gave up his bed in his private cabin, while he slept on a narrow bench which was much too short for him. Orde-Lees said of his generosity,

> He is a wonderful man. He looks after me himself with all the tender care of a trained nurse, which indeed he seems to be far more than merely my leader and master for the time being. He attends to me himself, making up the fire and making me a cup of tea during the night if I happen to say

> that I am thirsty, reading to me and always entertaining me with his wonderful conversation, making me forget my pain by joking with me continually, just as if I was a spoiled child. What sacrifices would I not make for such a leader as this.

As always, Shackleton never asked anyone to perform a task he would not undertake himself. He took his turn to scrub the floors, moved stores into the coal bunkers, joined teams on to the surrounding floes to hack off chunks of ice for water and went hunting for seals. Meanwhile, he kept the Australian photographer, Frank Hurley, busy, instructing him to fix lamps on 25-foot-high poles on either side of the ship to light up the decks. McNish, the ship's carpenter, was also instructed to build new sleeping cubicles for the men, with a blubber-fed stove pumping out heat to keep them warm.

It certainly wasn't all work. Leisure activities were also encouraged. Shackleton had again ensured that the ship's library was fully stocked, while theatrical shows, quiz sessions and plenty of poker, bridge and other competitive games were regularly laid on, not to mention the nightly sing-song after their meal.

One particularly successful form of entertainment involved the men cutting each other's hair. McNish described the newly shorn crew as looking like 'a lot of convicts'. Whenever possible, the men were also eager to escape the ship and exercise on the floes, where they played football, although a wire safety cordon was erected to help guide them back to the ship should a whiteout close in.

Recognizing the importance of food, Shackleton ensured his men always had full stomachs, leading to further contentment and reducing the risk of any bad temper. Much of their diet consisted of seal meat, of which there seemed a bountiful supply on the ice. The hunting teams certainly had no problem with slaughtering whole packs at a time, their blubber also saving on coal as fuel.

Yet Orde-Lees recognized that this might not always be the case and urged Shackleton to stockpile as much meat as possible. Shackleton disagreed, as he thought this sent out the wrong message, that the men might think they were trapped in the ice indefinitely. In his eyes, they had plenty of provisions to see them through the next few months and, as far as he was concerned, it was far more important that all of the men retained a sense of optimism.

Something many of the men enjoyed, and which was an important task, was exercising the dogs. Being cooped up on the ship was no good for them and they had to be kept in good shape for the overland crossing. However, by mid-April, many of the dogs had died due to various ailments. Shackleton could not afford a situation, as with the ponies on the *Nimrod*, where animal deaths contributed to the eventual failure to reach the pole. Now down to just fifty dogs, Wild, Crean, Hurley, Marston and McIlroy were all assigned to look after them. It was a job they came to relish, coming to see many of the animals as their friends. They particularly enjoyed hosting dog-team floe-races, which were bet upon with cigarette or chocolate wagers.

No matter Shackleton's attempts to remain buoyant, the men's mere survival was soon in question. After four months trapped in the ice, the ship had drifted over 650 miles. They were not only in uncharted waters, but in danger of being crushed. On deck, Shackleton joined the night-watchman and watched with concern as the massive pressure of converging floes forced chunks the size of double-decker buses out of the sea which then crashed back down, making a roaring sound like skyscrapers being demolished.

Some of the men saw this increasing movement of the

pack as an indication of an imminent break-up of the ice which might set the ship free, but Shackleton knew better. If the pressure could do this to massive blocks of ice, then the *Endurance*, made of oak, stood little chance. Growing increasingly concerned, he confided to Worsley that 'the ship can't live in this, Skipper'. Preparing for the worst, he ordered the three lifeboats to be lowered, just in case they needed to make an emergency escape.

Soon the pressure of the ice began to take its toll, slowly crushing the ship like an eggshell. Worsley described the sound of the ice pressing into her as 'like an enormous train with squeaky axles being shunted with much bumping and clattering', and 'it was making just the sort of sound you would expect a human being to utter if they were in fear of being murdered'. The floor linoleum began to crinkle at the edges, the walls slowly warped and the beams buckled and shook. 'The floes, with the force of millions of tons of moving ice behind them, were simply annihilating the ship,' Shackleton wrote. 'It was a sickening sensation to feel the decks breaking up under one's feet, the great beams bending and then snapping with a noise like heavy gun-fire.' No longer could Shackleton's optimism hide the truth from the men. The ship was in trouble.

By mid-October, the enormous pressure lifted the ship out of the water, tilting it at a 30-degree angle. For the men to eat at a table, or sit on chairs, they had to wedge their legs or arms against convenient ledges, although this didn't always stop their food from sliding off their plates. The ship then sprang several leaks, and while the men desperately manned the pumps, Shackleton could see that 'it was the beginning of the end'.

At 5 p.m. on 27 October, nine months after first becoming trapped in the ice, Shackleton gave the order to abandon ship. It was only a matter of time before the pressure cracked her in two and sent her spiralling into the abyss far below. Worsley tried to remonstrate, desperate not to lose his ship, and potentially his reputation, but he of course realized that, as Shackleton commented, 'what the ice gets, the ice keeps'.

With no time to waste, the men unloaded the boats, food, equipment and the dogs on to the ice. Though he recognized the gravity of the situation, Shackleton still gave the air of being totally in control. 'For most of the time,' Orde-Lees remembered, 'he stood on the upper deck holding on to the rigging smoking a cigarette with a serious somewhat unconcerned air.' Macklin also remembered that Shackleton showed no 'emotion, melodrama or excitement'.

As the men set up camp on the ice they watched on as their stricken ship was being destroyed. 'You could hear the ship being crushed up, the ice being ground into her, and you almost felt your own ribs were being crushed,' Commander Greenstreet recalled, 'and suddenly a light went on for a moment and then went out. It seemed the end of everything.'

Even if the ship should not sink, it was clearly beyond repair. Without it, the crew's position appeared hopeless. Thousands of miles away from civilization, they would now have to survive in tents on a drifting ice floe that could founder under pressure at any moment, sending them all to their deaths in the black waters below. There was no way to signal for help, and no one knew where they were. In any case, the prospects of a rescue party navigating through the ice in time to save them was extremely remote. Indeed, as far as the outside world was concerned, the expedition was all going to plan; they were not expected to be seen again for another five months, in March 1916. The twenty-eight men were therefore all alone, and in one of the worst predicaments Shackleton could have ever imagined.

30

For the previous nine months, Shackleton had clung to the fantasy that the expedition might still be able to go ahead. The loss of the *Endurance* now made this impossible. The only goal remaining was to somehow keep his men alive. 'I pray God,' he wrote, 'I can manage to get the whole party to civilization and then this part of the expedition will be over.'

It was a shattering blow to see his dreams once more collapse around him, but he daren't show any sign of weakness to the men. Beyond anything he had achieved in his life to date, this would be by far his greatest challenge.

The first night in their tents on the ice signalled just what difficulties lay ahead. In the darkness, the floe split apart, threatening to splinter the group, or send them crashing into the nearly open canals. At every fresh crack of the ice Shackleton raised the alarm by blowing into a whistle, at which the group worked feverishly to move the boats, tents and supplies out of harm's way. No one got much sleep, least of all Shackleton.

In the brief periods of calm the men still found it difficult to rest. Out on the ice in their unheated tents, they were exposed to the elements, particularly as there were only eighteen fur sleeping bags available. Lots were drawn, with those who missed out being handed inadequate woollen bags and consigned to shiver in the bitter cold. With only their noses protruding from their bags, any mucus quickly turned to ice, while every exhalation of breath formed a thick rime of frost as it met the cold air.

Shackleton could sense the gnawing anxiety among the men and knew the cold would not help their mood. At 5 a.m., with the ever-faithful Wild, he went from tent to tent, handing out hot cups of coffee, always with a smile, offering words of reassurance, downplaying the sinking of the ship by calmly saying, 'Now we'll go home.' Hussey remembered it as 'Simple, moving, optimistic and highly effective,' while Macklin said, 'It would be difficult to convey just what those words meant to us.' It was a masterclass in disaster management.

Shackleton knew that they needed to get to dry land as quickly as possible. They had drifted some 1,300 miles, and the nearest land was now Paulet Island, over 400 miles away. For this there were still some reasons to feel somewhat optimistic. In 1902, a Swedish expedition had been marooned on the island and had built a hut, leaving substantial stores behind. There was also plenty of wildlife, which would provide food, as well as blubber for fuel.

If they were to make such a distance, travelling light was imperative. Preparing the men, Shackleton ordered that they were all to only carry 2 lbs each, and to leave behind anything that wasn't of vital importance. To show that he was serious, he tossed his gold watch, as well as the Bible that had been presented to him by Queen Alexandra, into the snow. He clearly felt no need of God now, not that he was overly religious anyway. However, while the men packed, John Vincent retrieved the watch and hid it in his jacket. It was a decision he would later come to regret.

All vital stores and equipment were loaded on to two of the lifeboats at speed. Once full they were then hoisted on to the sledges, to be hauled by a mixture of dog and manpower. It was a daunting prospect and, with no room for passengers, Shackleton now had to make the most difficult call of all: the weakest dogs would have to be killed. Many found this order hard to take, even if they came to accept it. This was especially true of carpenter McNish, who was also ordered to kill his pet cat, Mrs Chippy.

On the morning of 30 October, the men began pushing and pulling the two sledge-mounted boats towards Paulet Island, hoping to use them once they found open water. It was back-breaking work, and made harder as the weight of the boats caused the sledges to sink deep into the many snowdrifts and ice. Grunting and groaning to move their loads, the men were soon exhausted. Calling a halt after just two hours, they found they had covered only one mile.

The next day they tried again, but still advanced only one more mile. It was obvious that reaching Paulet Island was a pipe dream. At the current rate it would take close to 400 days to reach their destination, and they only had rations for just fifty-six days. Moreover, with such taxing hauling, the men would also require far more calories than normal. The march to Paulet Island was therefore quietly abandoned.

If they were unable to reach dry land, then, for now at least, they needed to find a more solid floe on which to set up camp. Spotting one nearby, Shackleton christened it Ocean Camp. With the ship still close, the men were allowed to return to retrieve any of their belongings they had previously discarded. On boarding the stricken ship, the men waded through the ice-cold water and collected any personal items that might make them comfortable in the months ahead, as well as scavenging any items that might be of use to the camp. For instance, Wild salvaged the ship's wheelhouse, which, with the use of sails and tarpaulins, was fashioned into a storehouse on the ice. The ship's third lifeboat was also towed to Ocean Camp.

Hurley was particularly keen to collect his photographs, which were in the form of heavy glass plates. After some negotiation with Shackleton, who recognized it would be impossible to carry all 500 plates, Hurley was allowed to select, and keep, just 150 of the best, some of which you see in this book today. The 350 'rejects' were smashed, so that Hurley would avoid the temptation of carrying any more. Shackleton was growing increasingly impressed by Hurley; he thought him 'splendid' and added him, alongside Wild and Worsley, to the list of men he felt he could rely on for support and guidance.

Once camp was set up, and the men assigned into five tents, Shackleton faced some grim realities. The floe might have been more solid but they still faced the risk of drifting out to sea, or indeed of it collapsing all around them. This was clearly very much a temporary situation, yet what else could he do? Man-hauling over the pack had proven to be unsuccessful, and even if they could eventually get the boats into the water when the ice melted, they were not equipped for any long ocean voyage.

Making matters seem even more desperate was the sight of the *Endurance* finally sinking below the ice on 21 November. 'She's gone, boys,' Shackleton solemnly announced, as the water around her icy grave quickly froze over. With all the men watching, Worsley recorded that she 'put up the bravest fight that ever a ship had fought before yielding [to be] crushed by the remorseless pack'. Somewhat shellshocked, but putting on a brave face, Shackleton recorded in his diary, 'I cannot write about it.' If anyone had harboured any hopes of somehow making the *Endurance* suitable for sailing again, those hopes were now dashed.

Used to being a man of action, Shackleton had to, 'Put footsteps of courage into stirrups of patience.' No matter how much he wracked his brains for any sort of path to salvation, he came to realize that none was immediately forthcoming. McNish suggested using the wood from the three lifeboats to build a larger boat with which they could eventually set sail, but Shackleton decided against it. He wisely foresaw a period when summer loosened all the pack and escape would involve a dangerous period of movement between floes, with this boat being too heavy for the men to manhandle. McNish was instead put to work modifying the three boats, with sails rigged for all three and the wooden hull seams caulked against leaks, using Marston's oil paints, since no putty was available.

Realizing they could no longer wait for a solution to present itself, Shackleton, Crean, Hurley and Wild set out on a reconnaissance mission across the ice. The aim was to see if there might be an easier route to Paulet Island where they might also be able to utilize the lifeboats over small stretches of water, with the boats now named after the expedition's three principal backers: James Caird, Dudley Docker and Stancomb-Wills.

The prospects did not look good. At best, Shackleton thought they could travel 2–3 miles a day, but there didn't appear to be any open stretches of water. At this rate, it would take six months to reach Paulet Island. Despite these impossible odds, Shackleton had no option but to insist it could be done. It was all they had, and they needed some sort of hope.

On 22 December, the men enjoyed one final feast of meats, fruits, biscuits and hot drinks before the *James Caird* and the *Dudley Docker* were loaded and the men set off on their second man-haul attempt to reach Paulet Island. Alas, the conditions were no kinder than at the first attempt.

After the long weeks of inactivity at Ocean Camp, the extremely heavy loads and difficult terrain proved especially hard for some of the men. As they pulled the boats behind them, toiling in their harnesses like a gang of Egyptian slaves, the sun beat down on the slushy snow. Making little progress, Shackleton's aim of advancing 2–3 miles a day looked fanciful at best.

By 27 December, the men were at breaking point. McNish, who Shackleton had always been unsure about, and who still carried a grudge after being ordered to kill his cat, attempted a mutiny. Refusing to take another step, he cursed Shackleton for their predicament and for setting them on such a useless march, when they could have set sail on the boat he had offered to build. Apparently quoting the Ship's Articles, McNish claimed that now the ship had been lost, Shackleton had lost his authority over the men. At this moment, Shackleton might have wished he had learned the lesson from his experience with Marshall on *Nimrod* and had not taken with him any man he felt might turn out to be a bad egg.

As McNish ranted and raved, Shackleton sensed that the tired and discouraged men were in danger of being carried away with him. He knew he needed to stamp out the threat of mutiny immediately, or all would be lost. In contrast to the wild McNish, Shackleton calmly but forcefully read the Ship's Articles. No matter the loss of the ship, they made clear that he was still in charge. He told the men that they were also still being paid for every day they were on the ice. However, any man who chose to mutiny would not only not receive a penny when they reached port, they would be further punished by the full weight of the law.

The mutiny was crushed in an instant yet, just to make things clear, Shackleton took McNish to one side and warned him that if he were to repeat such behaviour, he would be shot. Dismayed and furious at this attempted treachery, Shackleton wrote of McNish in his diary: 'I shall never forget him in this time of strain and stress.' For now, Shackleton would brush the matter under the carpet for the good of the group, but it was an incident he would never forget nor forgive.

I have often found that when there is dissension in the group, the leader can often break the deadlock by getting up and personally tackling the task. Shackleton did exactly this, taking the lead the next day and hauling the boats harder than anyone, encouraging others to do the same. By the end of the day, the men had at last reached their target, covering 2.5 miles, but they were physically and mentally broken. In a week of marching they had advanced only 10 miles. 'The outlook was most unpromising,' Shackleton admitted, recognizing that the basis for McNish's outburst might have been well founded.

The next day Shackleton had no option but to abandon the march. This wasn't due only to the slow progress and the condition of the men, but the threat of the jagged ice doing irreparable damage to the bottoms of the boats, which they would surely need at some stage. Although Shackleton was a man of adventure, he was never reckless. As Worsley wrote, 'He was brave, the bravest man I have ever seen, but he was never foolhardy. When necessary he would undertake the most dangerous things and do so fearlessly; but always he would approach them in a thoughtful manner and perform them in the safest way.'

After setting up Patience Camp on another floe, Shackleton realized that any march over the ice in the current conditions was going to be impossible. All they could do was hope the drift took the floe close to Paulet Island, or wait for the ice to melt, then launch the boats into open water, setting off for the nearest dry land.

When reaching the North Pole in 1982, Charlie Burton and I found ourselves in a similarly perilous situation. Cut off from land and trapped on a floe, we were relying on it to drift south so we could reach our ship, which was unable to penetrate the pack. This was nerve-wracking, to say the least. At one point the floe split in two and threatened to separate us from our tent; it also began to melt, leaving many places 5 feet deep in ice-blue water. We also faced the menace of being trapped on the floe with polar bears who were sniffing us out as their next meal. When one bear showed particular signs of evil intent, we had no choice but to shoot it in one paw. After seventy days on the floe, 80 per cent of it was covered in slush or water, and killer whales were circling us. With the ship still 17 miles away, and our floe in a desperate state, we had no choice but to man-haul our way through the pack, where bergs were being tossed around in the waves like beachballs. I can barely describe our utter joy when we finally caught sight of the distant masts of the *Benjamin Bowring*. I jumped in the air and waved like a madman, until I could see Ginny on the deck. It was ten years since she had come up with the idea for our first surface circumpolar expedition around the Earth, and at last we had done it.

Shackleton's decision to be patient and wait for the floe to drift and reach open water was therefore a wise one. However, rations were down to just forty-two days, and soon the seals, which had been providing additional meat, had disappeared. He might at this point have rued his earlier decision not to stock up, in my view a reckless and unnecessary one.

The mood in the camp soon turned black, with Orde-Lees describing it as 'distinctly depressing'. There was now no real work to be done and the men had long since tired of any 'entertainment' to keep them occupied. 'Jimmy' James wrote, 'The worst thing is having time to kill.' Even the sparky Worsley reflected the downcast mood in his normally effervescent diary: 'It is to be hoped we can soon be "up and doing" something however little to aid our escape from this white, interminable prison where the mind's energies and abilities of all of us are atrophying and where we are resting and wasting our lives away while the whole world is at War and we know nothing of how it goes.' The dour and critical McNish, for now at least, wisely opted to keep his own counsel, instead spewing his spleen in his diary, calling some of the party 'useless' and criticizing others for having 'never done a day's work in this world and don't intend to as long as they can act the Parasite on somebody else'.

Minor issues, such as Orde-Lees's snoring, termed by Worsley as a 'nasal trombone', and Clark's relentless sniffing, threatened to spiral into serious disagreements. Green became concerned that some of the men had started to go 'crackers', with one putting a copper wire around his belly, threatening to walk home across the ice. For once, even Shackleton struggled to put on a brave face, writing, 'I am rather tired. I suppose it is the strain.'

A man in the prime of his health would have struggled in such conditions but Shackleton was far from this. On occasion, he required assistance from Wild to get out of his sleeping bag. When he eventually allowed the doctors to examine him, he again refused to allow them to listen to his heart, almost as if he knew the true nature of his illness. One thing was for certain, there was to be more stress and strain ahead.

A decision now had to be made about the fate of the remaining dogs. They would not be required to cross the ice, and they would not be able to travel in the boats. Furthermore, they were also eating from the dwindling stores of food. With a heavy heart, Shackleton decreed that they would all have to be killed. The grim task fell to an upset Wild, who wrote, 'I have known many men I would rather shoot than the worst of the dogs.' Their meat would serve to feed the men but some refused to eat it on principle, while Wild reported their steaks were the 'nastiest I have tasted, and the toughest'. Despite this, it was still the best meat on offer for the foreseeable future.

Rations now dwindled ever further, with Shackleton writing, 'we all feel that we could eat twice as much as we get'. He decided they would now have to break into the sledging rations, which he had been holding in reserve for any arduous treks.

Almost every day, groups of men returned to Ocean Camp to retrieve anything that might be of use, including the *Stancomb-Wills* lifeboat. This was to prove highly fortuitous as just twenty-four hours later a stretch of open water appeared, cutting the men off from any further forays to their former camp. They had retrieved the *Stancomb-Wills* just in time.

As the water became a little warmer, the ice on the floe gradually turned to slush, soaking the men's sleeping bags as they slept and carrying the very real threat that the ice could collapse beneath them at any time. The floe had melted from 1 square mile across to little more than 200 yards. The entire region was now in motion, slowly swirling and eddying. I well remember the spectacular noise and vibrations once the ice gets on the move. The fracture and pressure sounds are varied, but the most awe-inspiring are the booming and crunching. Like the pipes and drums of an approaching enemy horde, the distant rumble and crunch of invading ice floes approaching and growing louder hour by hour are impossible to ignore, and to sleep through. And yet the melting and moving ice and subsequent open waters might also be the men's salvation.

Thanks to the drift, by the end of February Paulet Island

was now just 100 miles away. Shackleton felt they would soon have to make their move. He therefore ordered the boats to be prepared and the men to sleep fully clothed, ready to set off at a moment's notice.

On 23 March, Shackleton looked through the mist in astonishment. He saw land the first time in sixteen months. Quickly consulting a map, it appeared to be the mountain-tops of Joinville Island, over 60 miles to the west, lying adjacent to Paulet Island. Yet a jigsaw of broken ice blocked their path. 'It might have been 600 miles for all the chance we had of reaching it by sledging across the broken ice,' Shackleton wrote. Soon after, the floe sailed past Paulet Island, leaving the prospect of reaching dry land ever more remote.

The floe was now becoming far too unstable to camp on for much longer. On one occasion it cracked in two, separating the men from the boats, before miraculously coming back together, allowing the men to reclaim them. Time was clearly running out. In a state of desperation, Shackleton consulted the maps. Over 100 miles to the north he spotted two tiny specks of land: Elephant and Clarence islands. Elephant Island was named after elephant seals by American sealers who landed there in the 1820s. A chunk of inhospitable rock in the middle of the vast Southern Ocean, it was frequently battered by hurricane-force winds and covered in ice. Shelter would also be hard to find and it was far from any shipping route.

The next option was King George Island, part of a chain of islands that would allow them to reach Deception Island, which was known to hold stores for shipwrecked mariners. It was 200 miles away, but was by far their best chance, particularly as whalers were active in the waters and they might be saved. If all else failed, Deception Island had on it a small wooden church. Shackleton proposed dismantling it and turning it into a boat in which they could sail back to South Georgia.

For the first time in months, Shackleton felt a surge of optimism. The odds were stacked against them, but at least there was something to aim for, a chance of salvation, no matter how small.

By 9 April, he could wait no more. After 156 days on the floe, it was now dangerously unstable and so small there was barely any room for the camp. With open water now all around them, it was now or never. At 2 p.m. the men jumped on board the pre-loaded lifeboats. Taking great care and using good psychology for the challenging voyage ahead, Shackleton had chosen which characters would best be grouped together and which kept apart. On the *James Caird*, Shackleton took with him the two most pessimistic men, Vincent and McNish, to keep them from spreading gloom and fear. Frank Worsley captained the *Dudley Docker*, while Hubert Hudson, navigating officer of the *Endurance*, skippered the *Stancomb-Wills*.

Splashing the three little boats into the floe-side network of open canals, they set off into the notorious Weddell Sea, which over a year previously had brought the far larger and more robust *Endurance* to her doom. Now, they were hoping to survive the same savage seas in just three lifeboats, travelling hundreds of miles with no cover from the elements. It was to be the ultimate journey from hell.

31

As Worsley tried to navigate a path from the helm of the *Dudley Docker*, Shackleton stood tall on the *James Caird*, looking out towards the dark horizon and praying that salvation lay ahead. 'I do not think I had ever quite so keenly felt the anxiety which belongs to the leadership,' he confided.

In the unrelenting darkness of the Antarctic winter, the conditions were even worse than they might have imagined. The sea churned with huge ice chunks, forever crashing together, with the men fearing they would be squashed between them. Unprotected, they were battered by the chilling winds, freezing sleet and fog, and sprayed relentlessly by the ice-cold sea. The threat of killer whales thrusting through the water and capsizing a boat was also ever-present.

The possibility of crossing the 200 miles to King George Island soon looked extremely unlikely. Shackleton realized that their best bet was instead to aim for Clarence or Elephant Island, just 60 miles away. They were, however, at the mercy of the wind. If its direction was unkind, even if they rowed with all their might in the opposite direction, they could be blown into the open water, with no other dry land for hundreds of miles. Still, it was the only choice they had.

After an arduous day rowing, the men found they still could not rest. That first night, they hauled the boats on to a giant passing floe, hoping to set up a temporary camp and get some sleep, no doubt believing it to be far safer than staying in the water. However, as the men slept, Shackleton was checking in with the night watchman when, directly beneath one of the tents, the floe cracked in two. Screaming for the men to get out before they plummeted into the ocean, he raced to the tent, only to find sailor Ernie Holness thrashing around in the water, still in his sleeping bag. Grasping the bag with his fingertips, Shackleton gritted his teeth, and using all his might hauled Holness to safety.

As Holness sat terrified and shivering, the gap where he had fallen just seconds before slammed shut. If Shackleton had acted a second or so later, Holness would surely have been killed. But there was still more drama to come. The floe quickly split again, this time casting Shackleton adrift, sailing into the darkness by himself, with no provisions or equipment. Wild was, thankfully, on hand and immediately launched the *Stancomb-Wills*, frantically rowing to the drifting floe and pulling Shackleton on board.

When dawn eventually broke, they wasted no time hurrying off the remnants of the floe and on to the boats. Soon after, it appeared they had met with some good fortune. The pack ice had disappeared altogether and they were finally out into the open sea. For so long they had prayed for this, but it was to prove a false blessing. Without the ice they were now pounded by great, rolling, breaking waves. With the heavily laden boats sitting low in the water, there was real concern that they might be overwhelmed and sink. It would be nothing less than suicide to continue, so Shackleton ordered the boats to return to the ice and set up camp, in the hope that the exhausted men would get some rest while Shackleton refined his strategy.

Again, there was to be no respite. The fierce winds drove the floe further back into the pack, and the risk of them being smashed by the enormous passing bergs made it impossible to launch the boats. All they could do was wait for a safe path to open up when the pack opened further. They were truly between the devil and the deep blue sea.

As the ice disintegrated all around them, Shackleton ordered the men on to the boats, yet as they all did so the ice broke beneath his and Wild's feet. Grasping at an ice ledge, the water lapping at their feet, they hauled themselves on. From now on, there would be no stopping for camp until they made it to dry land. They just had to hope the boats didn't sink in the ferocious swells.

The men now rowed even harder, not wanting to spend any longer in the water than was absolutely necessary. Yet dwindling rations meant that this extra exertion could not be compensated for, and some of the men almost passed out. Hot milk was regularly provided to keep them going, but it was not enough.

That night, the men slept in their boats, in the shadow of a giant iceberg, which at least provided some shelter from the bitter wind. Having no cover, and in soaking clothes, they huddled together for warmth. Hypothermia was just a whisker away.

When the boats started to rock in the darkness, some of the men awoke, believing the movement to be the result of strong waves. Yet when they opened their eyes, they saw that the ocean was flat. Then they heard the sound of hissing and were sprayed with water. They were being circled by a pod of killer whales. If one crashed down on to a boat, it would capsize. Once more, the men passed a sleepless night.

On 12 April, Worsley, catching sight of the sun, positioned his sextant to work out their position. It was not good. Despite all their hard work with their oars, the currents had left them no closer to the islands than when they had first set off, still 60 miles away from the nearest land.

That night was one of the worst they had faced on the ocean. In appalling conditions, and with no hot drinks available, the already freezing men were battered by barrage after barrage of water pouring into the boats. By the morning, they were covered in a film of frost, with thick ice hanging off their beards. Their clothes were so frozen that Wild described them as like 'wearing a coat of armour'.

Already starving, and wary of hypothermia, frostbite was a real concern, with most of the men having developed it somewhere. A breakfast of frozen dog food certainly didn't lift their spirits, with many of the men too sick to eat anyway. Riddled with diarrhoea, they defecated over the sides of the boats.

The lack of drinking water didn't help. In the open ocean, with no ice around them, there were no chunks to melt. The men's tongues subsequently swelled in their dry mouths, making it difficult to swallow food, despite their hunger. Wild believed that at least half the party were 'insane . . . simply helpless and hopeless'. On the *Stancomb-Wills*, four of the eight men collapsed, their feet turning a deathly shade of white. Perce Blackborow, the stowaway who had become cook's assistant, was in particularly bad shape. If his feet weren't warmed soon, he would lose his toes. Meanwhile, thick layers of ice rendered the oars too slippery to handle. Many of the men's frost-damaged fingers struggled to grip them in any event.

Facing another day of rowing in the cold ocean, unsure if they were making any progress, a sense of hopelessness descended. Shackleton, who, up until now, had been setting an example, was becoming a little frayed. His booming voice failed him and he could now only hoarsely whisper orders to Wild or Hurley. However, there was one moment that saw him raise his voice. When he saw that Vincent had his gold watch, which he had tossed into the snow months before, Shackleton grabbed it from his hands, and shouted, 'No, by God, you shall not have it,' and threw it into the sea.

Just as things seemed to be descending into chaos and

hopelessness, there came some hope. On 14 April, Clarence and Elephant islands came into view. Although still 40 miles away, this was the closest they had come: an encouraging sign that they were at last making headway. Shackleton tried to cheer everyone up by shouting to navigator Worsley that they should reach their island goal the following day. But Worsley dismissed the prospects, feeling it would take far longer. Shackleton was silently furious with him. He had only said as much to help boost morale. And soon after, morale was to take another shattering blow.

That night, Wild said the conditions were 'The worst I have ever known.' In temperatures of minus twenty degrees, a thunderous storm tossed the boats like toys on the sea and the swirling gale hurled wave upon wave of freezing water on to the men. Such was the force of the wind, the *Dudley Docker* was separated from the other boats and carried off into the darkness. It appeared all on board were lost.

The next day, the men were unable to muster much enthusiasm when the cliffs of Elephant Island came into view, mourning the loss of those on the *Dudley Docker*. Yet this was no time to mourn. A narrow beach on the island's eastern cape had been spotted, and all the surviving men needed their wits about them if they were to make landfall safely.

Approaching land, Shackleton suddenly caught sight of the *Dudley Docker* heading their way. All the men were safe, if battered and wet. They had survived the storm and found their way to the island at the very moment Shackleton and his men were set to land. It was the best news Shackleton had received in quite some time.

Soon after, the boats were brought ashore, and Blackborow was given the honour of being the first person to set foot on Elephant Island. Although such was the condition of his frostbitten toes, he collapsed and had to be helped to his feet. The rest of the men followed, putting their feet on solid land for the first time since 5 December 1914. Some frantically slaked their thirst with stream water, others staggered about like zombies, or manically laughed to themselves, appearing to Shackleton to be 'off their heads'. The actions of Green were particularly worrying. Seemingly in the grip of a breakdown, he launched himself at a pack of seals with an ice axe, according to Orde Lees, displaying 'all the primitive savagery of a child killing flies'. Reginald James wrote of the scene, 'Most people were I think in a semi-hysterical condition and hardly knew whether to laugh or cry. We did not know, until it was released, what a strain the last few days had been.'

The island was nothing more than a speck of rock in the cavernous ocean, but it was the safest haven they had been in for over a year. For now, Shackleton could rejoice at what he had achieved, but he knew this solution could only be temporary. Unless they were spotted by a passing whaler – rare in these waters anyway, let alone out of season – they would at some point have to undertake another journey to reach civilization. In the meantime, they had to get acquainted with the mysterious island they found themselves on.

32

There was to be no immediate rest for Shackleton. The shore where they had landed was just 100 feet wide and the cliff face behind was stained in water marks. It was a death trap. If they camped there, they would be swamped by the tide.

No alternative campsite could be found in the close vicinity, so Shackleton sent Wild, Crean, Vincent, McCarthy and Marston on a scouting mission. Boarding the *Stancomb-Wills*, they sailed 7 miles along the coast and soon found what Wild referred to as 'paradise'. This might have been stretching things, but at the very least he had found what appeared to be a safe spit of land. Just 100 yards long and 40 yards wide, it backed on to a 1,000-foot cliff and appeared to offer a ready supply of seals and penguins to hunt, with fresh water available from nearby glaciers. In the circumstances, it was all they could have asked for.

However, getting everyone to their new home, christened Cape Wild, was not so easy. While it was just 7 miles away, the boats now faced a fierce wind which threatened to blow them further out to sea. For over six hours the exhausted men once more toiled, rowing furiously against the wind, determined not to lose their chance when they were so close. The effort took its toll. When they arrived, in darkness, Lewis Rickinson, the chief engineer, suffered a heart attack. Luckily, he survived.

The boats were quickly unloaded, and they then built a shelter with rocks as walls and the upside-down *Dudley Docker* and *Stancomb-Wills* utilized as a roof, while tent materials were used to keep out the bitter winds and snow. The twenty-eight men were soon packed inside, desperate for warmth, and sleep, at long last.

Early the next morning they were awoken by hurricane-force winds. Tents were ripped from the ground, and personal possessions blown into the sea, including clothes and cooking pans. When trying to retrieve them, the men were almost blown off their feet.

Many of them realized that they would not last long in such conditions. But the nearest habitable landfalls were Port Stanley on the Falklands and Cape Horn at the tip of South America, both around 600 miles away. Elephant Island was not on any known shipping routes and there was no way they could signal for help. Worsley summed up their perilous situation best, stating, 'The world was as completely cut off from us as though we had come from another planet.'

Some of the men gave up hope. They had already been through so much, and now this appeared to be as good as it was going to get. Many believed they would die on the island and struggled to find the motivation to rise from their sleeping bags in the morning. Soon, a general sense of despair swept through the camp.

Shackleton recognized that their situation was only temporary. Somehow, they had to reach civilization and raise the alarm. The thought of getting back into a boat, however, on a journey that promised to be even longer and more hazardous than the one that had brought them here was unappealing in the extreme. And the odds of surviving it were slim indeed. Nevertheless, it was the only hope they had. Shackleton knew that, without any plan, some of the men would spiral into a dark depression and some would go totally mad, thus, he said, 'the conclusion was forced upon me'.

He realized that the best course of action would be for

him and a small party to lead a rescue mission while the others stayed on the island. However, he was well aware that their small boat would not survive the strong winds and currents to the Falklands or Cape Horn.

Consulting the map, he decided that their best chance of success was to return to South Georgia, from where the *Endurance* had first set out. This might have been over 800 miles away, but they would have the advantage of the westerly wind, rather than having to fight against it. There was, however, one distinct disadvantage. Such was the size and strength of the monster waves in this area, whaling skippers had christened it the 'great graveyard'. No small boat was known to have survived in its waters. Even the ever-optimistic Shackleton understood the odds would be against them, writing, 'The perils of the journey were extreme.' But if he was to make such an attempt, there wasn't a second to lose. By May, the pack ice would have closed off all available routes and they would face spending winter on the island. Indeed, Shackleton could already see pack ice on the horizon, creeping closer every day.

For such an arduous mission Shackleton would need to select his men wisely. Worsley was a natural choice. His experience at sea and brilliance as a navigator would be essential. Crean had also proven himself to be a more than capable sailor on the way to Elephant Island, as had McCarthy, described by Worsley as an 'irrepressible optimist'. To the surprise of some, Vincent was also selected. While he was a decent sailor, he was also thought to be a bully, pessimistic and light-fingered, not exactly the attributes Shackleton was looking for. However, Shackleton may well have felt that he would be a troublemaker if left on the island. On the journey, he could at least keep an eye on him. Perhaps Shackleton's most surprising selection was McNish, whom Shackleton had previously threatened to shoot. Yet this was a highly practical choice. The *Caird* was destined to take a battering, and repairs would be vital. The carpenter was therefore the best man for the job.

He would, however, have to leave some of his best men behind, in order to lead the dispirited group on the island and ensure they stuck together. Wild was selected for this especially important task, and the doctors McIlroy and Macklin would also stay on to look after the sick men, particularly Blackborow, whose frostbitten toes were now beyond saving.

A brief spell of hope swept across the camp as Shackleton made his final preparations. No matter that he had spelt out the dangers of the journey to everyone, and the slim likelihood of success, most still put their faith in the Boss. So far, he had not put them wrong and had displayed quite incredible powers of recovery and survival. If there was anyone on earth they wanted for such a mission, it was Sir Ernest Shackleton.

For the brutal journey ahead, McNish set to work preparing the *James Caird*. Utilizing whatever he could lay his hands on, he increased her hull height by 10 inches and stretched canvas over the top for shelter. Any areas that might leak were plugged with a concoction of oil paint, lamp wick and seal blood, while he installed the mast from the *Stancomb-Wills* along the keel, from bow to stern, as a measure to prevent the rough seas from cracking the *James Caird* in two. For ballast, rocks and shingle were also loaded on board.

By 24 April, the boat was ready to set sail. Saying his goodbyes to the men who would be left behind on Elephant Island, Shackleton keenly felt the weight of responsibility. They were all relying on him. This was his expedition, and he was determined to get every single one of them home. 'Skipper, if anything happens to me while those fellows are waiting for me, I shall feel like a murderer,' he told Worsley.

Taking Wild aside, Shackleton went through the worst-case scenario. If the *James Caird* was lost and no rescue ship had appeared by spring, Wild should attempt to reach Deception Island in the *Dudley Docker.* Resting his hand on Wild's shoulder and looking him squarely in the eye, Shackleton told him, 'I have every confidence in you and always had.' It was some consolation to him that the men remaining at Elephant Island would be in capable hands.

Showing at least some optimism, Shackleton also ensured that his book and lecture tour contracts would be fulfilled if he should be killed. Orde-Lees and Hurley were assigned to deal with the book and Wild and Hurley would undertake the lecture tours. Hurley would also be solely responsible for the expedition photographs.

Shortly before departure, Shackleton handed a note to Wild, to be read if he were to perish in the Southern Ocean. It simply read: 'You can convey my love to my people and say I tried my best.'

Moments later, Shackleton and his crew boarded the *James Caird*, cheered on their way by the men watching from the shore. 'She is our only hope,' Wordie wrote. Some of the men were in tears, recognizing that their fate lay in this mission's tiny chances of success.

After waving his goodbyes to the men as they disappeared from sight, Shackleton turned to the grey and desolate ocean ahead. They had a nightmare 800-mile journey in front of them, with just enough food and fresh water to last a month. Turning to Worsley, he said, 'We've had some great adventures together, Skipper, but this is the greatest of all. This time it really is do or die, as they say in the story-books.' This would turn out to be something of an understatement.

33

Straight ahead, and blocking the *James Caird*'s path was the constantly encroaching pack ice. Trying to row their way through the maze, the men soon realized it would be impossible to pass. The only available route was northwards, going around the pack before picking up the westerly winds they hoped would blow them all the way to South Georgia. This added more distance and time but, given the pack ice, it was unavoidable.

Yet just getting around the ice was still a major issue. A strong gale kept blowing them back in the opposite direction, while cross currents tossed the heavily laden boat around, flooding it with water. Worsley reminded Shackleton that he had warned that with the ton of rocks on board for ballast it was lying too low in the water. But it was not the time for recriminations. Clinging on for dear life, most of the men, including Shackleton, were so seasick they fouled the boat with their vomit.

Rolling, pitching and tumbling in the sea, they never had a moment to rest or gather their senses. Time was of the essence and they were making little progress. Worsley wrote, 'This was disheartening. We were as much drenched as ever by seas sweeping over us; and without the feeling that we were conquering distance.'

Thankfully, by the third day at sea, the wind had changed, allowing the *James Caird* to make its way around the pack and set course for South Georgia. Things began to look more promising, and they quickly covered over 80 miles.

With somewhat calmer weather, the men now settled into

a routine. Three men worked four hours at a time, one controlling the tiller, one the set of the sail, and the third bailing water continuously, then they switched shifts. The men not on shift were supposed to get some rest, but this was all but impossible. Freezing water continued to flood the deck, or trickle from the canvas above and down their necks as they tried to sleep. The frozen, wet sleeping bags also filled their nostrils with the smell of vomit and damp, and the moulting reindeer hair tickled their noses and lodged in their throats.

The only thing that could counter all this misery was mealtimes. The designated cook, Crean, would wedge the Primus stove between his feet in order to keep the contents from spilling while boiling the compressed sledging ration of beef protein, lard, oatmeal, sugar and salt. When the contents were bubbling hot, Crean would shout the magic word, 'HOOSH!' At this, the men would shoot their pots around the stove for Crean to pour each half a pound of the porridge-like mixture. Downing the scalding-hot contents as quickly as they could, the men burnt the roof of their mouths and their throats, yet did not flinch an inch. They more than welcomed the 'glorious heat' coursing through their cold and tired bodies, giving them a new lease of life as they moved closer to South Georgia and salvation.

As they entered the waters of Drake Passage, known as the 'great graveyard', towering waves, in some cases over 100 feet from breaking crest to roaring hollow, moving at 50 mph, slammed into the boat. The impact of a solid wall of water at such a speed had the potential to crush the *James Caird* and send it to the bottom of the ocean.

Again, the men had to cling on to the boat, and it was impossible for them to stay dry or to get any sleep. 'A thousand times it seemed as if the *James Caird* must be engulfed,' Shackleton wrote, 'but the boat lived.' Amidst all of this, the

Irishman, McCarthy, somehow kept his chirpy good humour, as Worsley recalled: 'I relieved him at the helm, seas pouring down our necks, one came right over us and I felt like swearing but just kept it back and he informed me with a cheerful grin "It's a foine day, sorr."'

The danger of the ferocious waves was in time replaced by sinking temperatures. A thick layer of parasitic ice crept along the canvas decking and the sails, leaving the boat top heavy, slowing her down and threatening to capsize her. To combat this new menace, they each took turns at the hazardous task of crawling over the madly tossing deck and hacking with one hand at the 15-inch layer of ice. Even wearing gloves, they could manage only a few minutes of this before having to squirm backwards into the bowels to try to unfreeze. When a thin layer of ice appeared on the surface of the water, the oars constantly froze to it, causing more problems than solutions. In the end Shackleton ordered all but two of them to be thrown overboard. The sails also had to be regularly taken down so that the ice could be removed, slowing their progress yet more.

Every man on board now had frostbite. Their feet had turned white and become numb, while such was their lack of sensation that one man didn't even realize his big toe had a pin stuck in it. Without a doctor, there was the real possibility that crude amputations would have to be carried out. To combat this Shackleton ensured that hot drinks were served regularly, and checked on the health of each man, always devising, according to Worsley, 'some means of easing their hardships'.

The first week at sea had been an ordeal to match anything that they had faced before. Thankfully, the waves settled and, more importantly, the sun finally shone. While the men put out their sodden sleeping bags, clothes and shoes to dry, Worsley positioned his sextant to try to work out their position. Until now, he had been using the position of the stars for guidance or, when that failed, the direction of the wind. Even with both, it was still incredibly difficult to take an accurate reading, with the boat relentlessly rocking back and forth. Worsley knew that his life, and that of his two dozen colleagues on Elephant Island, depended on his navigational expertise. Now he prayed that they were still on track for South Georgia. Any deviance from their course could be disastrous. If they missed the island, the current and wind would make it impossible to turn back, sending them out into the vast ocean and towards certain death.

With bated breath, the men watched as he made his calculations. They were at 56 degrees 13' S, 45 degrees 38' W – halfway to South Georgia. They were still on course, but then the ocean struck back.

On 5 May, Shackleton looked out to the horizon at midnight and saw what appeared to be a clear line of sky. On closer examination, he realized that it wasn't the sky at all. It was a tidal wave thundering towards them at relentless speed. 'During twenty-six years' experience of the ocean in all its moods,' Shackleton wrote, 'I have never seen a wave so gigantic.'

Raising the alarm, the men grabbed what they could to steady themselves as the wave smashed into them. 'There was a roaring of water around and above us . . .' Worsley remembered. 'The wave that had struck us was so sudden and enormous that I have since come to the conclusion that it may have been caused by the capsizing of some great iceberg unseen and unheard by us in the darkness and the heavy gale.'

Though the boat was still structurally intact, as the men wiped the salt water from their eyes, they saw it was engulfed in water, and sinking fast. Frantically, they grabbed whatever they could, bailing out pail after pail of water, until the *James Caird* was out of danger.

However, it was then discovered that their last water cask had cracked, allowing sea water to seep in. Little fresh water now remained. Shackleton had no choice but to impose strict rations on his already ailing men, just half a pint a day. 'Thirst took possession of us,' he wrote, of the ever-worsening situation, with frostbite, sores and boils already hitting each man hard.

Worsley and Crean now came to the fore. Remaining undaunted, and relishing the challenge, they welcomed the opportunity to prove they could take anything the ocean threw at them. Even in the midst of a wild storm, Crean would give as good as he got, happily singing old Irish songs with a smile on his face, daring the storm to do its worst. Worsley, meanwhile, was proving himself to be not just an expert navigator but a captain of rare calibre. Never shaken, he remained composed and purposeful, issuing orders with clarity, despite bedlam all around him. Shackleton had chosen his men well. They all complemented his leadership style and maximized their chances of success.

On the morning of 8 May, their fifteenth day at sea, lumps of kelp were spotted floating in the sea, and upon them two shags. Worsley turned and grinned. Shags never flew more than 15 miles from land, he told the men. South Georgia could not be not far away. Moments later, McCarthy cried out, 'LAND HO!'

The six men forgot their sore, soaked, thirsty state and experienced a once-in-a-lifetime sense of relief and happiness as they spotted cliffs ahead. It was South Georgia. However, the cliffs ahead were on the southern shore of the island, while they were looking to sail to the north, where the whaling stations at Stromness Bay would welcome them. Yet with Vincent and McNish particularly struggling, and with Shackleton not wanting to risk any currents or winds blowing them off course and away from the island for good, he gave the order to land the *James Caird* as soon as possible.

Aiming for King Haakon Bay, an inlet on the south-west corner of the island, they were thwarted by rocky reefs and high waves that threatened to sink the boat. To land in such conditions would be reckless. They had come so far, and now, with the prize in sight, Shackleton didn't want to take any unnecessary risks. His decision to spend another night on the boat, and try again in the morning, might not have been popular, but the men understood what was at stake. It was not just their own lives they had to preserve at all costs, but also those of their comrades back on Elephant Island. 'It would have been madness to attempt a landing,' Worsley admitted, 'in the dark, with a heavy sea, on a beach we had never seen and which had never been properly charted.'

After a restless night of excitement, battling with their vicious thirst, the men found at dawn it was still impossible to land. Somehow the conditions had worsened, with the wind and waves threatening to toss the *James Caird* on to the rocks. The decision was therefore made to head for Annenkov Island, 20 miles to the east.

The attempted journey was as challenging as anything they had yet faced. Hurricane-force winds threatened to tear the boat apart, the wood shuddering under its mighty force. At the same time, wave after wave thudded against the sides like a demolition ball, leaving the ship to buckle under the pressure. Unable to take any more punishment, the bow planks were forced open, sending more water into the ailing craft and forcing the men to bail like maniacs to stay afloat.

By nightfall they were no closer to landing and Shackleton was struggling to remain optimistic. 'The chance of surviving the night seemed small,' he recalled. 'I think most of us felt that the end was very near.' Worsley pessimistically added,

'What a pity. We have made this great boat journey and nobody will ever know.'

Just as it seemed the stricken vessel could take no more, at 9 p.m. the storm suddenly eased, the waves subsiding to a manageable swell. If it had lasted just a few hours longer, all might very well have been lost. Indeed, such was the might of the storm that it had sunk the *Argos*, a 500-ton steamer, in the same waters. Her cargo of coal was lost, as well as all her crew. It was a miracle the *James Caird* survived.

The storm might have subsided, but as dawn broke violent winds continued to slow their progress, holding them at the mouth of the bay for over four hours. They were within touching distance but yet still in severe danger. Thirst was now a real issue, and spending another night at sea wasn't an option. Somehow, they had to land, and with only two oars. It was time to go for broke.

Taking their frost-damaged hands to the remaining oars, the men rowed with all their might, trying to break through the wind and, in the fading light, find somewhere they could land. As they weaved a path through the jutting reef, Shackleton caught sight of a small cove. Urging the men on, each of them straining against the current, they finally felt a thud beneath their feet. The *James Caird* had run ashore.

Clambering out, the men staggered to a nearby pool of fresh water, falling to their knees and downing every last drop. Somehow, they had made it to South Georgia. However, the bay had not been their original destination, and for good reason. The nearest whaling station was on the other side of the island, and between them and safety lay an uncharted 3, 000-foot-high mountain range.

34

After a fitful night's sleep huddled inside a nook at the base of the cliff face, the men awoke to face the next stage of their mission. They had reached South Georgia, but safety, and rescue, was still far from assured.

Warming their hands over a fire made from driftwood, while devouring a breakfast of chunks of albatross, Shackleton went through their options. Ultimately, their goal was to get help at the whaling stations at Husvik, Stromness or Grytviken, all on the northern side of the island. By boat, this was 150 miles away. Assessing his men, Shackleton was not sure if McCarthy, McNish or Vincent could withstand another sea journey. Yet even if Worsley, Crean and Shackleton set off to get help, he was also not convinced the battered *James Caird* would make it. Not only had storm after storm pushed her to breaking point, but upon landing the rudder had also been lost. There was now no way to direct the craft.

The only option was to travel across the island by foot, a journey of just 30 miles. Yet this in itself was daunting. No one had ever before crossed the interior spine of South Georgia. To the naked eye, it looked like a series of mountain ranges, but who was to say what they would find. And they were totally unprepared for such a journey. Some of the men, such as McNish, McCarthy and Vincent, could now barely stand, and they also had no mountaineering equipment. If they needed to climb they would have to rely on their hands, most ridden by frostbite and blisters.

During my travels in the north as part of the Trans Globe

Expedition, Charlie and I faced a similar predicament. We had sailed for our 2,000 miles in an open boat through the North-West Passage to our destination, Ellesmere Island, but found ourselves on the wrong side. We had no choice but to cross over the inland range of snowbound mountains and glaciers on foot, a journey of 150 miles. We were already exhausted, but at least we could radio Ginny, who was able to fly us in some suitable mountain- and glacier-trekking equipment and provisions. Even so, the journey truly put us to the test. Charlie slipped on the ice and split his head open, and by the time we reached the other side our feet were swollen like balloons, our heels red raw and bloody, and our fingertips badly frost nipped. Without suitable equipment, it is very unlikely we would have succeeded.

Shackleton recognized there was little choice but to proceed, but with half the crew in no fit state for such an endeavour, the final leg of the rescue journey would lie in the hands of Shackleton himself, Worsley and Crean. He was, however, worried about leaving the three remaining men behind in the cove, which was in a precarious position. Backing on to a steep cliff face, there was nowhere for the men to escape if the tide came in or conditions worsened. The boat was also less than seaworthy, so they would be unable to move away from danger.

However, after a few days' rest, as McCarthy was staring out into the water, the lost rudder, with the entire ocean to choose from, suddenly washed up at his feet. At this stroke of good fortune Shackleton announced that they would all get back into the *Caird* and move to a safer shelter, likely to be found at King Haakon Bay, 8 miles away. He also hoped the mountain range interior might be a little easier to cross from there.

There was some trepidation about getting back in the

water, but at least the weather was far kinder. With no storms on the horizon, and the ocean calm, they took advantage of the westerly wind and sailed with ease to the bay, landing on a pebble beach flanked by snow-capped mountains. 'A great day,' Shackleton wrote, thankful for the – for once – painless journey.

With seals aplenty, there would be no shortage of food for those who remained behind. However, the bay was more open to the elements than the cove, so the *James Caird* was turned upside down on rocks to provide shelter. This saw the camp christened Peggotty Camp, after the family who lived under a boat in Dickens's *David Copperfield*.

Now Shackleton turned his full attention to the climb ahead. The whaling stations were just 30 miles away. He felt that in good conditions, the crossing should take them only a night and a day. To travel lightly, Shackleton decided they would go without sleeping bags or tents. All they would carry with them was three days' rations, a stove, matches, binoculars, two compasses and 50 feet of rope.

This, I believe, was the most reckless of all Shackleton's actions. It may have been taken in the interests of speed, but they could have spent a little time working out a way of lashing their tents and sleeping bags to their backs without greatly slowing their progress, while greatly increasing their overall chances of survival.

Yet just as they were making final preparations, the weather turned, forcing the men to shelter under the boat. Mid-winter was now upon them, and Shackleton worried that relentless storms would make it impossible to set off. And what if a storm erupted while on the crossing, when they were without sleeping bags or a tent? The ice-covered mountains were no doubt difficult to traverse at the best of times, let alone in foul weather. However, Shackleton could find no other solution. Once more, he had no choice but to be patient and pray there would be a break in the weather.

In the interminable wait, McNish made good use of the time. Extracting nails from the *James Caird*, he knocked them into the soles of the men's boots for crampons and made walking sticks out of lumps of driftwood. He also recommended they take his adze with them. These innovations would be invaluable.

Without warning, in the early hours of 19 May, the storm stopped. Shackleton quickly rose to his feet and ordered the men to make final preparations. After a hot meal, they made their way out of the *James Caird* 'hut' at 2 a.m., and shook hands with those who would remain behind, promising that they would return with help in just a few days. Saying their goodbyes, they then set off into the darkness on what they hoped would be the final part of their mission.

Thanks to McNish's crampons, and the walking sticks, the men made short work of the first climb, swiftly climbing a few thousand feet to reach the top of a steep slope. Under the moonlight, they could finally see what lay before them. Weaving their way through the snow-capped peaks were fields of ice and hazardous broken ground. It was no worse than expected, although neither was it better. It was merely confirmation that the next forty-eight hours were going to offer yet another formidable challenge.

The moonlight soon faded away, to be replaced by a thick mist that obscured their path. Walking tentatively onwards, unable to see much further than what was beneath their feet, Shackleton grabbed the other two men. They were on the edge of a precipice. One more step and they would have tumbled to their deaths.

In the interests of safety, they now roped together, walking in single file, with Shackleton, as ever, leading the way into the unknown. From the back, Worsley continued to act as navigator, trying to plot a course in the darkness. However, after a few hours of this, they found that they would need to turn back and retrace their steps. They had merely made a loop, making their way to Possession Bay, next to King Haakon Bay, and were once more staring at the ocean.

Frustrated, they moved back into the interior, stopping for some hot food to lift their spirits while Worsley reconsidered their route. To his eyes, the right-hand pass looked the easiest climb, which was certainly welcome news to the men. Yet in their condition, and with limited equipment, any climb was to prove a challenge.

The cold air and higher altitude, coupled with treading over the uneven terrain, quickly took its toll, requiring the men to rest every twenty minutes. 'We would throw ourselves flat on our backs with our legs and arms extended,' Worsley remembered, 'and draw in big gulps of air so as to get our wind again.'

Poor Worsley suffered more than the others. Shackleton and Crean at least had experience of marching in treacherous conditions in their attempts to reach the pole. They had learned, when their bodies were screaming for them to stop, how to harness their psychological strength in order to go on. Worsley had more than proved his endurance on the expedition so far, but this was a different challenge altogether. Noticing his struggles, Shackleton and Crean urged him on, and he certainly didn't skimp on effort.

Dawn brought welcome light as they hacked steps into a slope, tentatively scaling a mountain for three hours, only to find their route forward was blocked. At the top, they had hoped there would be a slope leading in the desired direction but instead there was a precipitous drop. 'I looked down a sheer precipice,' Shackleton wrote, 'to a chaos of crumpled ice 1,500 feet below. There was no way down for us.' Twice more, they had to turn back, with Shackleton becoming increasingly worried that the journey was going to take far longer than he had envisaged. He might now have regretted not taking any sleeping bags or a tent with them.

At each new ridgeline peak they saw mile upon mile of towering mountains stretching before them. 'We were in a solitude never before broken by man,' Worsley wrote, recognizing the scale of the task ahead. The only sound they could hear was the unnerving, thunderous roars of great avalanches and calving glaciers.

Continuing onwards, a 200-foot-wide chasm now obstructed their path, which Worsley said was large enough to contain 'two battleships'. There was no way directly across, and going back would not be an option, so on a narrow ledge, the men gripped the rock face and cautiously made their way around the edge, their numb and frostbitten hands offering the most tenuous of grips.

They could at least be thankful for the weather. It was still well below freezing, and the wind cut through their ragged clothes, but they had not had to contend with any storms. If any inclement weather were to strike now, it would be disastrous.

In the late-afternoon twilight, after climbing 4,500 feet to the top of another mountain, they found that the path ahead was obscured by another bout of thick mist. Unable to see ahead, Shackleton knew it would be foolhardy to continue into the darkness, but it was freezing cold at the top of the pass. At night, temperatures could fall to well below minus 30 degrees and, without any shelter, it would have been unbearable. 'I don't like it at all,' Shackleton said. 'We shall freeze if we wait here until the moon rises.'

There were deadly hazards no matter what Shackleton

chose to do, but he quickly made up his mind: 'We've got to take a risk.' He would rather die moving forwards than freeze to death waiting at the top of the mountain.

Using McNish's adze, Shackleton cut steps into the mountain for them to descend. In the darkness, they made just 100 yards in thirty minutes. At this pace, they could yet die from the cold. They had to move faster. Shackleton had only one solution. 'We'll slide,' he told the men, planning to coil the rope together to use as a toboggan. Worsley thought the idea was 'an almost impossible project', and even Shackleton admitted it was 'a devil of a risk'.

Short of any other option, Shackleton sat up front, with Worsley and Crean behind, wrapping their feet around the person in front of them. Bracing themselves for a moment, knowing they could all fall to their deaths or be battered against any number of unseen rocks, Shackleton pushed off with his feet. 'I was never more scared in my life than for the first thirty seconds,' Worsley wrote. 'The speed was terrific. I think we all gasped at that hair-raising shoot into darkness.' Shrieking aloud as they dropped, they picked up a fearful pace before coming to a stop. They had travelled at least 1,500 feet in just a few seconds. Slapping each other on the back, they laughed with relief. It was a miracle, and already they had so many to be thankful for. Shackleton, despite his lack of religion, had no choice but to credit his luck to a higher presence: 'I know that during that long and racking march of thirty-six hours over the unnamed mountains and glaciers of South Georgia, it seemed to me often that we were four, not three.'

At 6 p.m., with dusk turning to nightfall, they ate a hot meal, joining with Crean in old Irish songs, before returning to the march. They had been on the move for almost twenty-four hours, and had not slept a wink.

In the very early hours, hoping they were at least close to Stromness Bay, they came across a huge glacier. This was concerning. On their previous visit to Stromness Bay, no glaciers had been visible. They had clearly turned in the wrong direction. 'This disappointment was severe,' Shackleton wrote. The already tired men again turned back, wondering when this torture would ever end.

By 5 a.m. the men had exhausted all their reserves. They were shattered, mentally and physically. Stopping for a quick rest, Crean and Worsley drifted off into a deep sleep, and Shackleton had to fight the urge to join them. The temptation to rest, even just for a moment, must have been overwhelming, but, as he wrote, 'sleep under such conditions merges with death'. Somehow, they had to keep going.

Despite this, he allowed Crean and Worsley five more minutes of rest and, in yet another stroke of leadership genius, when he woke the men he told them they had slept soundly for half an hour. The psychological trick worked.

The somewhat refreshed men shortly began to recognize the slopes around them. They were similar to those surrounding Stromness Bay. Then, in the distance, they heard the faint sound of whistles. This could only be coming from the boats at the whaling station. 'Never had any of us heard sweeter music,' Shackleton recalled. Salvation was within sight and, down to their last rations, just in the nick of time.

After a quick snack to power them for the final leg of their journey, they set off for the whaling station. Yet the going wasn't easy. There were still steep icy slopes and deep snow to negotiate. But the smell of freshly slaughtered seals urged them on, a sign that civilization was within touching distance.

At 1.30 p.m. they had climbed the last coastal ridge and gained a clear sight of the Stromness whaling station. It was the first time they had seen other human beings in eighteen months. Excitedly, they waved and shouted at the matchstick figures 2,500 feet below, only for the wind to drown out their cries.

Now, moving ever quicker, they descended the final slope towards Stromness, their excitement bubbling over, before Shackleton came to a stop. Before them was a waterfall with a sheer drop of 15–30 feet. There was no way around it, but there was no going back either.

Tying a rope around a boulder, one by one, they abseiled down, the cold water splashing their faces and soaking their clothes. All were soon safely at the bottom and the whaling station now just 1.5 miles away.

Staggering on in their threadbare clothes, the excited men came across two children. When Shackleton asked where they could find the station manager's office the children were so scared they fled. Soon after, an elderly man did the same, shocked at the sight of these wretched creatures, who, with their blackened faces, straggly beards and strong stench, barely looked human. Worsley could not blame them, admitting that they were 'Ragged, filthy and evil-smelling: hair and beards long and matted with soot and blubber; unwashed for three months, and no bath or change of clothing for seven months.'

As they entered Stromness at 4 p.m. more and more people fled from the staggering scarecrows, as if trying to escape a plague. Finally, Matthias Anderson, the station foreman, stopped, and Shackleton hoarsely whispered that they wanted to see the manager. Leading the men to the office, Anderson left the suspicious characters outside while he spoke to the manager, Thoralf Sørlle. 'There are three funny-looking men outside,' he told Sørlle, 'they say they know you.' Indeed, Shackleton and Sørlle had met in 1914, before the *Endurance* had left the island.

Emerging from his office, Sørlle looked the ragged

creatures up and down. He had certainly never met them before. 'Who the hell are you?' he shouted, ready to run these vagabonds out of town. 'My name is Shackleton,' came the reply. Looking him in the eye, Sørlle saw the same twinkle he had seen eighteen months before. The world had given Ernest Shackleton up for dead, but now here he was, somehow still clinging to life. But as Shackleton was soon to learn, the world he had left behind was now very different to the one he now re-entered.

35

Shackleton, Crean and Worsley had survived one of the most remarkable tests of survival known to man. Eighteen months after leaving the shores of South Georgia, they had returned, having lost their ship, camped on melting and cracking ice floes, escaped killer whales, spent weeks in the deadly Weddell Sea being pummelled by giant waves and icebergs and climbed uncharted mountains without proper equipment. They looked like they had been through hell, but they had lived to tell their remarkable tale. And only just. Hours after their arrival, a snowstorm struck. Had it hit during their climb, it would have spelt the end. Subsequent records at the whaling station indicate that, the remainder of that winter, there was no period when humans without tent or polar clothing could have survived such a journey.

In the sanctuary of Sørlle's house, the men eagerly took up the offer of a bath, washing away the sweat and dirt that had accumulated over the last eighteen months. Worsley waxed ecstatic about his bath in his diary, which reminded me of the truly wonderful feeling of my own first hot bath after ninety-four days of polar man-hauling without washing or changing clothing. I could have had a hot shower the previous day, but refused, rather than dilute the true perfection of *real* submersion of all the nerve ends, which remains my definition of sheer sensory bliss.

Sørlle's steward then gave them fresh clothes and served up coffee, bread, jam and cakes. It was a slice of heaven. While preparations were underway to find boats to save the men on the other side of the island, not to mention still stranded on Elephant Island, the conversation turned to what they had missed while they had been away, with Shackleton asking, 'Tell me, when was the war over?'

On the eve of their departure, in August 1914, Britain had just announced it was at war with Germany, with many predicting that it would be over by Christmas of that year. Shackleton and the men had therefore believed that the guns had long since been silenced. Now Sørlle told them that the war was still raging, and getting worse by the day. 'Europe is mad,' Sørlle exclaimed. 'Millions are being killed.' His talk of trench warfare, poisonous gas, tanks and U-boats, all reaping carnage, chilled the men to their core.

'No other civilised men could have been as blankly ignorant of world-shaking events,' Shackleton remarked. It seemed impossible that while he and his men had been stranded on the ice, Europe was being torn asunder.

That night, Shackleton was unable to sleep. Beset with worry, he feared what the future held in store for his country, and his family. South Georgia had no cable head or wireless so he was unable to send a message home that he was alive and well. For now, however, he still had unfinished business in the south.

While Shackleton continued with his plans to rescue the men on Elephant Island, who had now been there almost a month, Worsley steamed his way to King Haakon Bay, on board the steam whaler *Samson*. After a few hours' sleep, the freshly shaven Worsley was soon walking up the beach where he had last left the men. On seeing him, McNish, McCarthy and Vincent did not at first recognize the clean shaven Worsley, and were so weak they could barely muster a cheer, but smiles still spread across their haggard faces. Their bitter ordeal was over. Towing the *James Caird*, the *Samson* took them back to Stromness, where they could at last rest.

Meanwhile, Sørlle was helping Shackleton and Crean find a ship, and crew, that would take them to Elephant Island. As word spread quickly, Shackleton was offered the services of the *Southern Sky*, a four-year-old steam whaler, which would be captained by Ingvar Thom, along with a full Norwegian crew. Shackleton was also beset with offers of donations for the journey. For this, he was extremely thankful, as the expedition had long since run out of money, and he didn't want to waste any more time having to beg and plead for funds.

With the Norwegians hard at work preparing the vessel, Shackleton and his comrades were thrown a party in the local saloon to celebrate their journey. In a packed room smelling of blubber and tobacco smoke, Shackleton took the opportunity to dust off his public-speaking skills. This was, however, a tough audience. The Norwegian whalers had seen all manner of things during their time at sea and were not easily impressed or inclined to give praise. When Shackleton had finished speaking there was silence before an older man rose to his feet, pointed at Shackleton and his men, and announced, 'These are men!' With this approval, the Norwegians all eagerly shook the hands of Shackleton, Crean and Worsley, now almost in awe. 'Coming from brother seamen, men of our own cloth and members of a great seafaring race like the Norwegians,' Worsley recalled, 'this was a wonderful tribute and one of which we all felt proud.'

The very next morning, Shackleton, Crean and Worsley set off for Elephant Island on the *Southern Sky*. Meanwhile, McNish, McCarthy and Vincent made their way home, where they would join the ongoing war effort. Tragically, McCarthy, whom Worsley had affectionately called 'a big, brave, smiling, golden hearted Merchant Service Jack', was killed at sea just six months later.

The chief concern for Shackleton in reaching Elephant Island was the winter pack ice, which any time now would cut off the men at Camp Wild for a year. He had known this could be an issue but hoped they could get to the island before it had surrounded it. Yet as soon as they encountered the first ice it was clear that the *Southern Sky* was not built to plough her way through. With just ten days' worth of coal on board, they had little time to try and navigate a path. Still struggling, and with 75 miles still to go, Shackleton made the decision to turn back and find a more suitable ship.

Rather than return to South Georgia, Shackleton directed the ship to Port Stanley in the Falkland Islands, some 500 miles to the north. There, he hoped not only would he find a more suitable selection of ships, he could also finally make contact with London.

Arriving on 31 May 1916, Shackleton sent a 2,000-word cable to the *Daily Chronicle*. Shackleton had long been presumed dead, so the news caused a sensation. The front-page headline the next day read 'Safe Arrival of Sir Ernest Shackleton at Falkland Islands'. Meanwhile, Shackleton also wrote to Emily, declaring he was indeed alive but had endured 'a year and a half of hell', promising to return home as soon as he had rescued every last one of his men.

This was not going to be easy. Finding a suitable vessel to break through the ice was proving far more difficult than he had thought. Not only were there none available in the Falklands, but on asking the Admiralty for assistance he was told that while they could send the *Discovery*, she would not arrive until October. Shackleton could not afford to wait that long. When he had left Elephant Island, many of the men were already in a pitiful state. If they had to wait another four to five months to be rescued, many could die.

In a 'fever of impatience', according to Worsley, Shackleton fired off cables to anyone he thought might be able to help. The Foreign Office soon cabled back. After scouring

South America for a suitable vessel, it had found the *Instituto de Pesca No. 1*, a steam trawler then based in Montevideo. At 280 tons, the ship was made for heavy seas and came with free provisions and crew. By 16 June, she was at Port Stanley, with Shackleton, Crean and Worsley on board, again making their way to Elephant Island.

And yet only some 20 miles from their destination, twenty-eight-year-old Lieutenant Ruperto Elichiribehety, who was in charge of the ship, ordered a retreat. He did not want to risk being trapped in the ice, when the boat was already struggling badly, with the engines failing and coal fast running out. 'It was a dreadful experience to get within so short a distance of our marooned shipmates and then fail to reach them,' Worsley recalled, with Shackleton left 'nearly heartbroken'.

Back at Port Stanley, Shackleton found that, in the interim, HMS *Glasgow* had arrived at port. The ship looked perfect for the rescue mission, and Captain John Luce was eager to assist. Contacting the Admiralty to gain its permission to set off for Elephant Island, Luce soon received a stern reply: 'Your telegram not approved.' With a war raging, the Admiralty could not afford to take the *Glasgow* out of action.

Increasingly frantic, Shackleton decided to cut his losses and head to Punta Arenas, Chile, with its thriving port and sizeable British community, hoping to find a ship there. Securing a berth on the mail ship *Orita*, Shackleton, Crean and Worsley arrived on 1 July 1916, where the British Consul, Charles Milward, understood just what they were up against. The fifty-seven-year-old had himself once been shipwrecked in the Antarctic and wanted to help in any way that he could. Introducing Shackleton to Allan MacDonald, the President of the British Association of Magallanes, a gentleman's club, a gathering of members was called for 9 July, where Shackleton could appeal for help.

In front of a large crowd, Shackleton regaled the many ex-pat Brits with his adventures, while also making a plea for assistance. Moved by the incredible story, a fund was quickly set up, with the Brits and wealthy Chilean businessmen, making contributions amounting to more than £2,000. Rather than beg for a vessel, Shackleton now had the money to rent one. He subsequently secured the services of the *Emma*, a 70-ton oak schooner. As part of the bargain, and to help save coal, the Chilean authorities also agreed to provide the steamer *Yelcho* to help tow the *Emma* towards the ice.

Under the command of the Chilean pilot León Aguirre Romero, she was soon at sea, but Shackleton's optimism was revealed to be misplaced. It seemed the *Emma* was no better equipped for the job than the *Instituto de Pesca No. 1*. This time with 100 miles to go, the auxiliary engine broke down, while the ship had already sustained heavy damage from the ice. There was a real fear that they could become trapped, not only putting themselves in danger but further delaying the rescue of the men on Elephant Island. To the relief of all on board, the engine was soon re-started but it was clear they could go no further. Just three days into the journey, the *Emma* was turning back to port, the situation more desperate than ever.

Shackleton had now had to abandon three rescue missions on three different ships. Meanwhile, the men had been stranded on Elephant Island for almost four months, and the pack enclosing their camp seemed impenetrable. 'He did not speak of the men on the island now,' Worsley wrote of Shackleton's despair. 'It was a silence more eloquent than words.'

His hair turning grey from the strain, Shackleton turned to whiskey to somehow dull the pain at what he saw as a monumental failure. While he might have survived, he believed that he had now lost the lives of most of his party. For the previous eighteen months his goal had been to save them all, but it now appeared he would not succeed. Rather than feeling like a hero, he felt like a villain. 'His days were bad,' Worsley said of this period. 'What his nights were like I can only imagine.' With Shackleton sinking his sorrows in ever more bottles of whiskey, Crean told him to moderate his drinking. There was still a chance, he said, however slim, that they could save the men.

It appeared that Shackleton's last and only hope was to wait for the arrival of the *Discovery*. She was soon to set sail from England, but the journey would take six weeks and Shackleton had been informed that Captain James Fairweather, a sixty-three-year-old Scot, would be in charge of the rescue mission. As such, Shackleton would be relegated to the sidelines. He was enraged. The men were his responsibility, and his alone. He knew the area, and the task at hand, better than anyone. There was no way he would step aside.

Furious and frustrated, Shackleton decided to try to find another ship. His options in Punta Arenas were, however, limited, especially as his funds had run low after the disaster with the *Emma*. The *Yelcho* was one of the few vessels that might be available, but the boat was clearly unsuitable. A rusting hulk with a suspect engine and boiler, she had also not been strengthened to cope with the ice. Moreover, she had a top speed of just 11 knots, far too slow to plough her way through. Despite all of this, Shackleton was so desperate he asked the Director General of the Chilean navy to release her. He did so, but on one condition: the boat was not to go near the ice. Shackleton agreed, knowing full well it was a promise he could not possibly keep.

On 25 August, Shackleton set sail once more. If he failed this time, he would have no choice but to step aside and let the *Discovery* take over. This only spurred him on, and they progressed quickly to within 60 miles of Elephant Island, with the pack nowhere to be seen. But then, around midnight on 29 August, floes of ice came into view, only to then disappear in a thick fog. To keep his promise, Shackleton should now turn back. Indeed, it seemed madness to continue into the pack with such an unsuitable ship, in low visibility. Nevertheless, somehow Shackleton managed to persuade Luis Pardo, a thirty-four-year-old pilot from the Chilean navy, to continue onwards.

With the ship at half speed, Shackleton manned the bridge, binoculars strapped to his face, squinting into the fog to try to map a path forward. Yet as they entered the pack the ice crumbled aside; it was not as solid as on previous missions. Ploughing onwards, scarcely able to believe their luck, by daybreak the mountains of Elephant Island appeared on the horizon.

Drawing ever closer, Shackleton lifted his binoculars and directed them towards where the men had last set up camp. There he saw tiny figures jumping wildly up and down, waving frantically in his direction. Squinting, he tried to count how many he could see, praying that he would find all twenty-three men still alive. 'There are only two, Skipper,' he began. 'No, four, I see six, eight . . . and at last . . . they are all there. They are all saved.' It was a moment of sheer exhilaration.

Moving purposefully towards the island, they were soon within 150 yards of the shore. Lowering a small boat into the water, Shackleton, Crean and four Chilean sailors clambered on board and made their way to the camp, which they had left in very different circumstances over 128 days before.

Standing upright, Shackleton could hear the sound of whooping and hollering from the shore, particularly when the men realized that their saviour was none other than the

Boss himself. He had kept his promise to them and achieved the impossible. Yelling to Wild, Shackleton cried, 'Are you all well?' to which his great friend replied, 'We are all well, Boss.'

There was no time for Shackleton to set foot on Elephant Island. At any moment, the vulnerable *Yelcho* could become trapped in the ice, so he needed to get the men on board as swiftly as possible. Over the course of the next hour, he took two trips to load the men on to the small boat, and then on to the *Yelcho*. Once everyone was on board, he could now see that 'some hands were in a rather bad way', particularly Blackborow, whose toes had been amputated by McIlroy. To save their strength, Shackleton directed them to rest below deck while the *Yelcho* headed north.

Still, they could not relax. The *Yelcho* once more needed to beat the ice, while sailing into the teeth of a furious gale. Luck was, thankfully, again on their side, and by God they had all earned it, as the pack again opened up to the plucky *Yelcho* and her crew, almost as if it knew they had suffered enough. At last, Shackleton now had the time to speak to Wild. Unsurprisingly, they had had a bleak time of it.

Shortly after Shackleton had left, food had been stolen, and Wild had to threaten to shoot any man found stealing. But with the 'Boss' gone, and hunger hitting hard, madness and anarchy gradually threatened to overrun the camp. Wild knew he needed to get a grip, and fast. Using Shackleton as his inspiration, he recognized that, more than anything, the men needed hope, so each day he told them to 'lash up and stow, boys, the Boss is coming today'. According to Orde-Lees, the men came to respect Wild's 'buoyant optimism' and sense of fairness, which allowed him to lead by example, rather than right. By earning the men's respect, he was able to keep a firm lid on things. It was an act of extraordinary leadership, for which Shackleton commended him.

However, Wild told Shackleton that if he had not arrived by 5 October, it had been decided that Wild would lead a five-man rescue team on the 200-mile trip to Deception Island on the *Dudley Docker.* Shackleton was relieved he had reached the men in time. There was no way the boat would have survived such a trip.

Now en route to Punta Arenas, Shackleton called for a quick stopover at Rio Seco, where he made a quick call to the governor. The *Yelcho*, he told him, was coming back to port, with all twenty-three men from Elephant Island on board. It was something that surely required a celebration.

Thus, when the *Yelcho* steamed into port on 3 September, the harbour was lined with virtually the entire population, cheering and waving, a brass band blaring 'God Save the King'. Despite the fact that the war still ongoing in Europe, even the German and Austrian ships in the harbour flew their flags to honour the returning men. No matter their differences, this was a heroic story that men from all over could appreciate. As Orde Lees said of the scene, 'this was no mean home coming'.

Back in civilization for the first time in twenty-two months, the men from Elephant Island posed for a photograph outside the city's Royal Hotel, still in their ragged and filthy clothes, as directed by their PR-savvy 'Boss'. Alongside them was a clean-shaven Shackleton, beaming proudly, his mission accomplished.

Writing to Emily later that day, Shackleton managed to capture all of his emotions: 'I have done it,' he triumphantly declared. 'Damn the Admiralty. I wonder who is responsible for their attitude to me. Not a life lost and we have been through Hell. Soon I will be home and then I will rest.' But before he could 'rest', what about the other half of his expedition team?

He still had to find out what had happened to the *Aurora* party, who had made their way to McMurdo Sound, unaware of Shackleton's travails on the other side of the continent. Shackleton had assumed that at the very least they were all safe and well, despite the expedition as a whole being a failure. But he was soon to find they were anything but, and while the rest of the *Endurance* party now made their way home to fight in the war, Shackleton was soon steaming his way to New Zealand on another rescue mission.

36

Since landing in South Georgia, Shackleton had heard only scattered reports that the *Aurora* party were marooned on the ice at McMurdo Sound. He was, however, given no reason to believe that Mackintosh and his crew were in any imminent danger. After all, they had stores for three or four years, the *Aurora* was back at port, and their location was known. No doubt they would soon be rescued, although he wasn't sure why they hadn't been already. In any event, his first priority at South Georgia had been saving his men on Elephant Island, who were in a far more precarious situation.

Learning that the *Aurora* party were still on the ice, Shackleton, along with Worsley, went to Panama, where they boarded a boat bound for New Zealand. He wrote to Emily, 'I did think that at last I was going to get home after all these last two years of strain and anxiety; but it cannot be helped.' He felt that the men on the *Aurora* were his responsibility and if any rescue attempt was still required, then he would lead it. But this time, as he set sail, he learned that things would be very different.

In his absence, there had been growing criticism of the expedition from the British, New Zealand and Australian authorities. The *Aurora* had clearly been underfunded and underprepared, and this had led to the men's lives being at risk. As Mawson said, the authorities were concerned that 'all that explorers in future will do is to raise enough money to get away to where they want to do their work, then call out to the Government to complete the job'. All fingers of blame pointed at Shackleton, and although he had saved the men on Elephant Island, the authorities now wanted him to have no part in arranging any mission to McMurdo Sound. Indeed, William Ferguson Massey, the New Zealand Prime Minister, warned that there was 'no reason' for Shackleton to visit.

Shackleton was outraged, but Mawson had no sympathy, claiming that his 'crooked dealings have brought it on himself'.

It transpired that not only was Shackleton to be left in the cold, he would not be allowed any input into choosing the man who would be in charge of the rescue. When he suggested that Stenhouse, the skipper of the *Aurora*'s first journey to the ice, should be considered, he was overruled. The cost of the rescue mission was around £20,000 (£850,000 today), and the authorities felt that as they were paying for it, they would decide who should lead the mission. Besides, Stenhouse was also being blamed for the party being stuck on the ice in the first place. The authorities subsequently appointed John King Davis, who had previously travelled on the *Nimrod* expedition as chief officer, and who Shackleton had initially asked to command the *Endurance*.

Having been hailed as a hero in Chile, the switch to Shackleton being portrayed as a villain was hard for him to take. Yet, still en route to New Zealand, there was little he could do, other than send a series of cables demanding that the *Aurora* not be placed under Davis's command. The authorities stood firm, tired of reckless expeditions and more concerned with the horrific war in Europe, which had seen the deaths of over 10,000 Australians and New Zealanders in Gallipoli alone. But Shackleton wasn't ready to quit just yet.

On 2 December, he and Worsley reached Wellington and immediately went to see Robert McNab, New Zealand's shipping minister. A master in emotional intelligence, Shackleton recognized that ranting or raving would not work. Instead, he laid on the charm and made McNab think again. Perhaps Shackleton could lead the *Aurora* to the ice, or at least be on board. A compromise was worked out. Davis would remain captain of the *Aurora*, but Shackleton would be on board as a 'supernumerary officer', supervising any land journeys that might be necessary. The authorities had someone they could trust in charge, while Shackleton would lead any rescue attempt, if required, once they reached land.

Assembling a crew of twenty-five men, Davis sailed the *Aurora* out of Port Chalmers on 20 December. While Shackleton had fought so hard to lead the mission, being relieved of command at least allowed him to relax a little. Happily trading stories with the crew, he also took the time to reminisce as he passed milestones from his past, including Mount Erebus, which hissed steam into the bitter chill from her volcanic top.

Shackleton was now forty-two. He had already written to Emily and told her, 'I don't suppose for a moment that the Antarctic will ever see me again.' He recognized that this would probably be his last chance to set his eyes on the ice, and he wanted to savour every moment. As Davis said of the journey, it was a 'voyage of the past'.

Approaching Cape Royds on 10 January, Shackleton could make out the hut which had served as a camp for the *Nimrod* eight years before and now served as camp for the trapped Ross Sea party. Having happily stayed in the background on board, now Shackleton went ashore, accompanied by two men from the *Aurora*, to lead the rescue of his men.

However, on reaching the old hut, they found it was deserted. Instead, there was a note stating that the men were now based at Scott's old hut at Cape Evans. Immediately setting off for Cape Evans, Shackleton and the two men from the *Aurora* soon caught sight of the men staggering towards them, filthy, haggard and wild, just as Shackleton himself had been just a few weeks before. Yet only seven of the ten men were present. Through the grime, Shackleton could make out Ernest Wild, Cope, Gaze, Jack, Joyce, Richards and Stevens, but where were Mackintosh, Hayward and Spencer-Smith? The horrific story of the men's experience at McMurdo Sound soon came to light when the seven were on the *Aurora*.

Having set out from Tasmania to the Ross Sea in December 1914, the men soon landed at Cape Royds, and anchored the *Aurora* to a floating glacier known as the Glacier Tongue. Yet as the men were unloading their stores, the anchor chain snapped, sweeping the ship away in embedded pack ice, along with vital food, clothes and gear. The ten men found themselves marooned, and did not have enough stores to survive. Making the most of an appalling situation, they ransacked Shackleton's old hut at Cape Royds, and then Scott's old camp at Cape Evans, for any supplies that might have been left behind by previous expeditions. It was slim pickings and it seemed all the men could do was focus on their own survival, rather than embark on their depot-laying mission. Indeed, by now, many of the dogs were already dead and much of the sledging equipment was still on the *Aurora*.

However, Mackintosh believed that Shackleton was relying on them. He had no idea that at this point the *Endurance* was already trapped in the ice on the other side of the continent and the traverse plans were therefore all but over. He wrote, 'Such setbacks and surprises where life and death are mingled so closely I have not experienced before. And it's hard to be existing with a sword of Damocles suspended over one as [Shackleton's] life – for the responsibility lies on my shoulders.' He subsequently told the others that, despite their perilous circumstances, they must somehow lay the depots. Shackleton was depending on them. All the men agreed.

Their aim was to lay five depots across the Barrier and up to the bottom of the Beardmore Glacier. To achieve this, they would need to man-haul over 1,000 miles, 15 miles a day, every day, with the help of just four dogs, and with inadequate sledges and supplies. Such a pace had been beyond Shackleton and the *Nimrod* team eight years previously, with the aid of ponies and sledges, and more rations. Unsurprisingly, they struggled badly, barely able to march more than 5 miles a day.

Tensions were soon rife, particularly between Joyce and 'One-Eye' Mackintosh, with Joyce commenting, 'I have never in my experience come across such an idiot in charge of men.' For all Shackleton's faults, in situations such as this he was peerless, and Mackintosh was to be found wanting, although he also had to counter some extraordinarily bad luck.

While Cope, Gaze and Jack had to abandon the march, Spencer-Smith bravely continued, despite being exhausted and unwell with the dreaded scurvy. As the illness took its toll, Spencer-Smith was left behind in a tent to recover while the remaining five men travelled onwards to Mount Hope to lay a depot.

On their return to Spencer-Smith's tent, all the men were now suffering from scurvy. It was clearly time to return to Hut Point. They had taken themselves as far as was humanly possible. Such was his condition, Spencer-Smith had to be carried on a sledge while Mackintosh was also unable to haul, leaving Joyce, Richards, Hayward and Ernest Wild to shoulder the extra burden. In appalling weather, as their fuel ran out and the food ran low, Hayward also became too weak to haul, making matters more desperate still. Such was Mackintosh's distress that he wrote a letter to his family in case of his demise.

On 7 March 1916, with 30 miles still to go to Hut Point, Mackintosh could go no further. The rest of the party left him behind, continuing onwards, promising to return. However, as they did so, on 9 March Spencer-Smith fell into a coma and died. He was thirty-two and had been ordained as a priest only five days before leaving London.

There was no time to mourn. The men needed to reach Hut Point quickly, with Hayward was only just hanging on. Struggling onwards, they finally reached Hut Point, barely alive. Urgently requiring food and rest, Hayward was left to recuperate in the hut while Joyce, Richards and Wild returned for Mackintosh on the Barrier, carrying with them seal steaks to help ward off scurvy.

They reached Mackintosh just in time, and he soon felt a little better after eating, allowing the four men to set off for Hut Point. However, they now found that miles of unstable ice had cut them off from Cape Evans and there was apparently no alternative route through. They had no choice but to wait for a more stable route to appear as the darkness of the Antarctic winter descended upon them.

Finally, on 8 May, after seven weeks of interminable waiting, Mackintosh could stand it no more. Despite no safe route being available, and it still being dark, he was determined to give it a go, with Hayward alongside him. Joyce warned them profusely that such an attempt was nothing more than suicide, but they would not listen. Staggering off into the darkness, the thin ice creaking under their feet, they would never be seen again.

The surviving men now all feared the worst. While they believed that the *Aurora* was lost to the Southern Ocean, they had put their faith in Shackleton's march across the continent being well underway. All being well, he should have reached them around 20 March, as he had planned. But there was no sight of him. The men presumed that he and his party were dead and that they were now all alone.

Since breaking its chain, the *Aurora*, and the crew who were left on board, had been swept out to sea and had travelled north for some 1,600 miles. When she was finally released from the ice pack in March 1916, her rudder had been smashed and she was inoperable. Thankfully, the wireless was still working, allowing operator Lionel Hooke to make contact with a New Zealand station, revealing their own situation, as well as that of the men marooned at Cape Evans. The Morse code messages were, however, brief and like a chain of Chinese whispers. These scant and scattered details were the ones that had eventually reached Shackleton on his arrival in South Georgia, and had now led him back to McMurdo Sound.

On hearing this story, Shackleton was heartbroken. Three of his men had lost their lives trying to lay depots for him, against unimaginable odds. In the process, they had set a world record of sledging for 1,215 geographical miles, and for spending nearly 200 days on the Ice Shelf. (It wasn't until 1993 that this record was broken, when Mike Stroud and I man-hauled an entirely unsupported distance of 1,350 geographical miles, 2,504 kms.)

At the news of Mackintosh and Hayward's demise, Shackleton sorrowfully shook his head. 'I wish to heavens that they had kept together,' he muttered under his breath, while directing praise to Joyce: 'your conduct on that long, trying southern journey – especially after Mackintosh broke down – ranks in my mind, and will, when the other men know of it, rank with the best deeds of Polar Exploration'.

Shackleton took full responsibility for the deaths, even though he was thousands of miles away at the time. His chaotic management of the *Aurora* had not helped matters, but the party had also encountered all manner of bad luck. Nevertheless, he wrote, 'My name is known as an explorer, as the leader of the Expedition. With the leader lies the praise or blame, and rightly so.'

Many of the men from the Ross Sea party also held Shackleton accountable and were barely able to hide their contempt for him. Even Joyce was somewhat critical, blaming Shackleton for crewing the *Aurora* with men who were 'only fit for drawing room parties', and for 'the playing of chances in equipping the ship in Australia'. Joyce would, however, calm down a little after learning of Shackleton and his team's horrific experiences, recognizing that he had 'nearly put up with as much hardship' as the Ross Sea team.

Shackleton was so distressed that the *Aurora*'s doctor, Frederick Middleton, examined him, finding him to be 'greatly affected' and 'very much disturbed', writing, 'Sir EHS is not at all well and I don't think he is in too fit a condition.'

This whole mess was entirely due to Shackleton's reckless impatience. His overall plan to cross Antarctica was bold and laudable, but too complex to succeed without step-by-step, meticulous preparations. This was a lesson when planning our own polar circumnavigation some fifty-eight years later, which was taught by the Royal Geographical Society Expedition Committee under its then president, Ernest's son, Eddie Shackleton.

As our one-time Norwegian rival Erling Kagge once wrote, 'Even a mouse can eat an elephant if it takes small enough bites.' So fixated had Shackleton been on his goal, and on beating his rivals, that he took far too many short cuts, putting all of his men's lives at risk. His leadership on the ice and subsequent rescue missions had been heroic, but he should never have put himself, or his men, in such a situation.

Rather than return to Britain triumphant, Shackleton now carried with him the guilt of the deaths of Spencer-Smith, Hayward and Mackintosh. It was the darkest moment of his career and, with it, the period of Antarctic exploration, known by many Shackleton biographers as the 'Heroic Age', was all but over.

37

Returning to New Zealand with the *Aurora* and his men, Shackleton was engulfed in bad publicity, as well as demands for long-outstanding payments. He knew he had to settle his affairs before things got seriously nasty, and he was extremely thankful that he had a fair amount of support in doing so.

His long-time Kiwi friend Leonard Tripp had worked feverishly in raising funds to meet debts, working out favourable deals with angry creditors and ensuring that Shackleton was off the hook for the cost of the relief expedition, which came to over £20,000. The sale of the *Aurora* for £10,000, three times above what Shackleton had paid for her, was a particularly brilliant piece of wheeling and dealing, with Tripp recognizing that all such boats were now in high demand, due to the war. This money also allowed Tripp to pay members of the Ross Sea party their wages in full.

Overdue payment was, however, to remain a bone of contention for many of the *Endurance* crew as, despite having promised to pay them for all the time they spent on the ice, Shackleton now said he was able to pay them only up until January 1915, when the *Endurance* was first trapped. This left a bitter taste in many mouths, with some turning their back on him. But there was simply no more money available. Shackleton tried to make amends in any way that he could, writing glowing letters of recommendation for the likes of Joyce and Crean and providing a free loan to seaman Walter How. He was also a constant visitor at James McIlroy's bedside after he was badly wounded in the war.

During his time in Australia and New Zealand, Shackleton set to work on his next book, *South*, which was again being ghostwritten by Edward Saunders. However, he was not overly enthusiastic about this book, for he would never make any money from it. The rights were owned by Sir Robert Lucas-Tooth, as a guarantee against a loan of £5,000 he had provided for the expedition.

Shackleton was also too raw to regurgitate everything he had just endured. Watching his dictation sessions with Saunders, Tripp recalled him struggling to find the words, tears brimming in his eyes, telling him, 'Tripp, you don't know what I have been through and I am going through it all again, and I can't do it.' Somehow, the book was completed, and eventually released in 1919. It remains in print to this day.

With everything tied up, as best it could be, and with all the men from both expeditions now home, Shackleton, rather than heading home to his family, set off for America, where he had arranged a lecture tour. *Endurance* was likely to be his last expedition, so he had to capitalize on his story, and raise funds, while he could. Yet, as always, Shackleton gave a large percentage away to charity, as well as to a trust fund set up for Mackintosh's wife.

Another reason for the trip to America may well have been that Shackleton realized that, once he returned home, his adventures would be over. He would surely be far too old, or sick, to ever again head south. And who would back him now, after everything that had happened? Tripp had made all this pretty clear to him. 'It seems to me that it will be impossible for you to do any exploring anyhow for many years,' he warned, continuing, 'It would be unwise for you ever to take on another expedition unless you not only had sufficient money to pay your way, if everything went alright, but you would have to have money in hand to provide for your accidents.'

Shackleton had missed his family, but he clearly did not miss domestic life. He knew that once he was home that would be the fate that awaited him, potentially for the rest of his life. By this time, he hadn't seen his family in three years and he was very fortunate that, in Emily, he had a wife who understood him. Indeed, Emily said to Tripp about her husband's reluctance to return home, 'I know it would bore Ernest to be here for any length of time.'

But Shackleton was far from happy on his American lecture tour. America entered the war soon after he landed there, and audiences stayed away, no longer captivated by the tales of an explorer. As he said after a show in Tacoma, 'There has been a mistake in the name TACOMA – the T and the A should come off and it should read "COMA".' A sense of sadness now pervaded Shackleton. The spark that had once enraptured audiences was dimming, as if he knew the end was near. Writing home, he confessed that he had 'grown very weary and lonely'. Now and again, he showed flashes of the old Shackleton, particularly at a packed Carnegie Hall, where the full house brought out the best in him. Sir Shane Leslie, who was in the audience that night, recalled, 'Shackleton was like a man who had discovered the Poles or the equator . . . we were all raised into a frenzy . . . it was the most exciting, tremendous meeting I have ever attended in the United States.' Sadly, such performances were now few and far between.

By the middle of May the tour was at an end and Shackleton could no longer avoid the inevitable. Boarding a boat in New York, he soon arrived in an England very different from the one he had left three years before. Unlike the fanfare he had received on returning after the success of *Nimrod*, this time there was no one to greet him. The country was preoccupied by war and the millions dying at the Front. It had no praise to lavish on a man who had seemingly escaped the fighting.

With no great greeting he instead went to Eastbourne, where his family were living in the quiet suburbs. Although he had done all he could to delay his return, he was still overjoyed to see them all again, especially his children. Raymond, now twelve, Cecily ten and Edward five, had very few memories of their father being at home and this was a chance to make up for lost time. However, it didn't take long for Shackleton to become frustrated. Without any sense of purpose and feeling that his best days and chances of further success were behind him, he was bereft. 'I only hope he will get something to do that will interest him,' Emily wrote, 'as he could never be happy in a quiet domestic life.'

The thunder of artillery from across the Channel, vibrating through the night, made him desperate to join the war effort, but at forty-three, he was too old for conscription and a desk job would not suit his buccaneering character, as he had previously found to his cost. What he really wanted was a chance to get stuck in and prove his worth once more, that he was not over the hill and still had plenty to offer.

Suddenly, an idea from the past sprung to mind. At the end of the Russo-Japanese War he had looked into the prospect of shipping Russian soldiers home. Once again, Russian soldiers needed transportation, especially through the winter snow. Surely, he thought, there was no man more suitable than him? But the Foreign Office did not take the bait and left him grounded in Eastbourne.

Wracking his brain for how else he might be of service, and once again sail to exotic locations on the high seas, he conjured up another idea. South American countries such as Argentina and Chile had remained neutral in the war, and Britain was desperate to get them on side. Shackleton, who remained revered in the region, floated the idea that he be sent on a propaganda tour to help with this. His idea was approved and, as hastily as ever, he said his goodbyes to his family and was once more off to seek adventure, writing happily to Emily en route, 'I have the ball at my feet now, I must kick hard but carefully. All my love to you. Bless you – Your own old "Micky".'

Yet by April 1918 Shackleton was already back home, his mission having made very little impact in South America, certainly in regard to gaining support for the war. He was now incredibly frustrated. Most of those who had served on the *Endurance* and the *Aurora* were fighting on the Front in some capacity, but he, the fearless leader who met any difficulties head on, was having to sit the war out.

Shackleton soon fell back on bad habits. Turning to drink, he once more sought company in other women. Rosalind Chetwynd, now known by her London stage name Rosa Lynd, came back on the scene, as did Belle Donaldson, after whom he had named Mount Donaldson. Yet Shackleton no longer carried the allure of the buccaneering man of adventure. While he told Donaldson that he might again go south, it now seemed more of a sad pipe dream, that of a man whose better days were behind him searching for a purpose.

However, in the summer of 1918, a potential opportunity popped up, in the Arctic Circle, of all places. The Northern Exploration Company (NEC) wanted Shackleton to head to Spitsbergen, to lead the search for gold, coal and other valuable minerals. At the same time, he was to investigate any German presence on the island. It all seemed too good to be true. A far-off adventure, promising danger and intrigue, as well as the opportunity to make his fortune at long last. To his delight, Wild and McIlroy agreed to join him.

While Shackleton did set sail in mid-July, he only made it

as far as the Norwegian port of Tromsø, where a message from the War Office was waiting for him, ordering him to immediately return to London. This might have been a blessing in disguise. It was already clear that Shackleton was unwell. He was suffering from his familiar pattern of breathing problems and allowed McIlroy to examine him, but not to listen to his heart. There was, however, little doubt that this was the real issue.

Back in London, there was at least a mission lined up for Shackleton. Post-revolution Russia had erupted into civil war. British interests at the key strategic ports of Murmansk and Archangel were now under threat, and the Germans were thought to be ready to take advantage of the turmoil. Shackleton was therefore tasked with heading to the region to protect the ports and assist the counter-revolutionary White Russians with supplies and equipment. This news certainly cheered him up.

On 22 July 1918, he was appointed as a temporary major in the army and sent on his way. Joining him were many members from his various expeditions, including Worsley, Stenhouse, Hussey, Macklin and Marshall. As Worsley wrote, 'The old gang was on the warpath!'

For an all-too-brief moment Shackleton was back at his best. His old friends beside him, he was off on another adventure, and had another chance to make his mark. Major General Maynard, his superior, soon fell under the Shackleton spell, saying of him:

> Shackleton had a great hold on the imagination of the men, who were keenly interested in hearing him tell of his past experiences, and ready to take to heart the lessons he inculcated. He proved a cheerful and amusing team member, and during the long, dark winter of 1918–19 his presence did

> much to keep us free from gloom and depression. He became almost at once a member of the happy family of which my Headquarters Staff consisted, taking and giving his full share of chaff, and pulling his weight in helping to provide amusements and entertainments for all ranks.

Alas Shackleton did not stay long in Russia. Just two weeks after his arrival, the First World War came to an end. Although the civil war in Russia was to continue, the British were no longer concerned by the German threat to their ports. In February 1919, Shackleton was on his way home, but he was not dispirited, as he had yet another business proposition that would apparently provide him with a fortune.

After just a brief time in Russia he felt he could capitalize on the post-war development around the ports of Archangel and Murmansk. By setting up a company to hold exclusive fishing, timber and mineral rights, he thought he could export these items all over the world. 'The possibilities of Russia,' he proclaimed, 'are beyond belief . . . It is not unlikely that she will lead the world in mineral production.' Once more, he tried to persuade the beleaguered Emily that he had hit the jackpot. 'The trading alone is worth £250,000 pa,' he wrote, 'so at last all is well.'

Nothing came of it. Shackleton had no money nor the business experience to get such an ambitious project off the ground, and those who had money didn't trust him with it. Once again, he was broke, unemployed and listless. Tripp recommended that 'the position of Consul would be in your line', but this was far too formal a role for Shackleton to consider. The only thing he could now do was lecture.

At least the lectures were at last something to shout about. They coincided with the publication of *South* as well as the release of Hurley's incredible footage and pictures from

Endurance. Thanks to this, Shackleton was soon 'packing them in' to the Philharmonic Hall in London, twice a day, six days a week, for five months. The swagger and charm were soon back, but he made little money. Incredibly, he still continued to give vast sums away to charities such as the Middlesex Hospital, ever mindful of the fanfare that would follow his generosity, while the rights to the book lectures and film were owned by others and would continue to be until Shackleton had paid off his various debts.

Soon enough the lectures wound down, playing to half-empty houses, and Shackleton found himself no better off than he was before. The future appeared bleak. Forced to rely on handouts from Janet Stancomb-Wills, he had proven he was not a man of business and was not suited for life behind a desk. Now the war was over, there was also no chance of military action. He readily admitted that all he was really good at was leading expeditions, and the more he thought about it, the more he believed that was all he had left.

Like the subject of one of his favourite poems, 'Ulysses' by Tennyson, he was the king who had returned home determined for more adventure, desperate to see out his final days 'sailing beyond the sunset and all the western stars until I die'. In a letter to Emily, he admitted as much, saying, 'I am no good at anything but being away in the wilds just with men. I feel I am no use to anyone unless I am outfacing the storm in wild lands.'

Despite the odds again being against him, in the spring of 1920 he began to look at mounting one last expedition, with some of his old pals brought back for the adventure. By hook or by crook, Shackleton was determined to return to the ice but, this time, with a twist.

PART FIVE

'Ah, what is the secret you're keeping to the
Southward beyond our ken?'

38

While Shackleton had failed to reach the South Pole, or cross the Antarctic continent, he had certainly seen enough of the south to last a lifetime. So, rather than give it one more go, he now decided to try his luck in the north.

His new goal was to reach the North Pole and to explore uncharted Arctic areas. He hoped, in doing so, to find missing Inuit tribes and to study magnetism. To do this, he estimated that he would need to raise £50,000. However, even at the height of his success, using his high-society connections to back him, and relying on his previous semi-success with *Nimrod*, this had still been difficult. Now, times, and Shackleton, had changed significantly, making it more difficult still.

In post-war Britain there was no longer any great excitement about such expeditions. Rather than cheering on such nationalistic conquests, it seemed the country was now undergoing a period of reflection. The death of Scott, coupled with Shackleton's most recent disaster, had also tarnished the once-great push to explore unknown regions. The country was mourning millions of deaths, and the idea of spending vast sums of money on Arctic explorations seemed foolish when it could be better spent on addressing post-war poverty.

In the past, Shackleton might have glided past such concerns with his charm, but the years had dimmed the lights. Smoking and drinking heavily, his business deals having all come to nothing, and with an uneasy relationship at home, a pervasive air of sadness now hung over him. It seemed that the determined young bull with a twinkle in his eye was but a distant memory.

The government also showed next to no interest in backing him, and neither did any of his previous donors, while the Royal Geographical Society, reluctant at the best of times, was now more focused on conquering Everest. Shackleton had hoped to persuade many of his old mates from *Discovery*, *Nimrod* and *Endurance* to join him for one final hurrah but, for many, the strains of the last expedition still cut deep. Some had no interest in ever putting their lives in Shackleton's hands again, while others simply wanted to avoid any further icy experiences. Many, such as Crean, admitted that they had grown too old for such adventures. They were now settled with their families, content with all they had achieved. Shackleton might have taken a moment to think likewise, but he knew that would never satisfy him.

In any event, Shackleton plunged headfirst into making plans, even though he faced the real prospect of being barred from the Arctic. In order to head north, he had to seek the permission of the Canadian government. However, the Canadians were less than enamoured with his plans, or indeed with him. He was far from a safe bet and the last thing they wanted was to have to pick up the tab for any rescue missions should things go wrong.

Just as it appeared things were grinding to a halt and Shackleton was facing living out his dotage in domestic boredom, an unexpected figure from the past rode to the rescue. John Quiller Rowett, a friend of Shackleton's at Dulwich College, had since made his fortune in rum, apparently cornering the world's market. Hearing that Shackleton was planning another expedition, he was only too happy to make a contribution. When Rowett learned that the expedition was still grossly underfunded, he offered to pick up the tab in full, putting £70,000 at his old friend's disposal.

Typically, with the money in his bank account, Shackleton did not think to wait for the Canadian government's approval. He instead set off to Norway, where he splashed out £11,000 on a four-year-old wooden ship called *Foca 1*, renaming her *Quest*. A hundred dogs were speedily ordered, as well as other supplies and equipment, including an electrically heated crow's nest, and an odograph to measure the ship's course. Perhaps the most extravagant purchase of all was an aircraft. He had no idea if the aircraft would work in the Arctic, but he intended to use it to spot new islands, reefs and any suitable sites for wireless relay and meteorological reporting stations.

Despite this lavish spending, by May 1921 the Canadian government had still not given its approval. Any attempt to push them along, with Shackleton even going so far as travelling to Canada to persuade them, fell on deaf ears. Finally, in the summer of 1921, he got an answer, but it was not one he was happy with.

The Canadian government was set to fight a general election and was not willing to back any venture to the Arctic that could blow up in its face. However, there was a sliver of hope. They told Shackleton that if they were to remain in power, they might give him permission to travel the following year. This was no good. Shackleton had a boat and supplies ready to depart and he couldn't face the prospect of another year at home. He needed to get away. Despite all of his preparations being for the north, he now decided to head back to the Antarctic.

Reaching the pole or crossing the Antarctic were ruled out, so at the last minute he conjured up a plan to circumnavigate the continent. Mapping a 2,000-mile stretch of uncharted territory in Enderby Land, he estimated it would take two years to complete. Always eager for any expedition to have the prospect of unearthing riches, he also hoped to search the South Atlantic for the lost treasure of the pirate Captain William Kidd, as well as for rich mineral deposits. To gain the approval of the relevant authorities, he also promised to study weather patterns and glacial formations in the region.

However, it was now clear that his motivation to discover new and exciting things was on the wane. When asked why he was continuing to explore, now approaching middle age and with the disasters of *Endurance* and *Aurora* still in mind, he merely answered, 'I am mad to get away,' or 'Because I like it and because it's my job.' Although he might again be heading south, the fire that had once lit his belly was on the way out. It was now more something to do than any great urge to prove himself by shattering some new geographic record.

When news broke that Shackleton was returning south, some old names from the past offered their services. These included the ever-faithful Wild, McIlroy, Hussey, Kerr, Macklin, McLeod, Green, Dell and Worsley, who would skipper the *Quest*. The Shackleton–Rowett Expedition was described by one of the more cynical members of the team as 'Shackleton's long, but not entirely selfish, joyride'. It was indeed more like an excuse for an old boys' reunion than a serious voyage of discovery.

As Shackleton had struggled to garner much media interest for his latest expedition, Emily, who had become a key figure in the Girl Guide and Boy Scout movements in Shackleton's absence, suggested running a competition for two boy scouts to join him. The idea was seized upon by the *Daily Mail*, and there were over 1,700 applicants, with Robert Baden-Powell personally selecting the lucky two: eighteen-year-old James Marr and seventeen-year-old Norman Mooney.

When the *Quest* arrived in London in mid-August, there

was time for the now obligatory audience with royalty to wish Shackleton well. King George V presented him with a silk Union Jack, while dowager Queen Alexandra, now seventy-seven, and always bewitched by Shackleton, also paid him a visit. Again, this seemed more a procession of ceremonies rather than there being any excitement at the prospect of the journey that lay ahead. It was almost as if he was a rock star going out on his last greatest-hits tour, the fan base having long since diminished.

On 17 September, the *Quest* was ready to set sail for the Antarctic. Yet some expressed concerns about Shackleton's health. He was clearly not the young buck he had once been. The weight gain, jowls, wrinkles and greying hair were evident, and he moved more slowly, owing to flat feet. Shackleton had even told Hussey, 'When we get on the ice, if my feet are bad I'll be able to get on the sledge and you can drive me along and you fellows can run alongside.' However, there were more worrying signs that all was not well.

Mill, who knew Shackleton better than most, thought his old friend looked jaded, his face showing signs 'of the wear and tear of his long years of unceasing hardship and toil'. Worsley agreed, saying, 'At the outset Shackleton did not appear to me to be physically the man he was when he led past expeditions.' None of this was surprising, of course, but those who knew Shackleton best were well aware that it was his heart that was the issue. Yet once again, while Shackleton reluctantly saw a specialist, he refused to let them listen to his heart. His health issues were instead put down to indigestion and rheumatism. Shackleton might well have known he wasn't anywhere near his best, but there was no way he was going to miss out.

Before setting off, Shackleton met Emily in Plymouth to say his goodbyes. He had previously told her, 'I think you are a wonderful girl and woman to have stood my erratic ways all these years.' Of that there could be no doubt. Once more she would be left to care for the family and manage the finances, not knowing when Shackleton would be home, or indeed if she would see him again. Perhaps she welcomed the peace, knowing that her husband was unhappy with domestic life. Or perhaps she hoped that one day Shackleton might achieve all of his dreams and could then return home and be happy. She prayed that at long last this supposed final expedition would give him all he wanted.

As the ship sailed into stormy open seas it became apparent that Shackleton had once more purchased a vessel that was unsuitable for the journey. At best, it moved at 5 knots, so rapidly fell behind schedule, while the boiler cracked, the engine shook and the ship sat so low in the water that her decks were regularly swamped.

Urgent repairs were required in Lisbon, which revealed that the stormy voyage had strained the *Quest*'s rigging and bent the engine's crankshaft out of alignment. When further repairs were then needed in Madeira, Shackleton wrote to Emily, 'Before I go into the ice they must be put right and if this takes too long I must do the islands only this year and early next year and hope that I may get sufficient results to justify the Expedition without having to go for any time into the ice.' After the repairs were made the ship still struggled on the voyage to Rio de Janeiro. Further inspection at the port showed that a month of further repairs were needed to make her seaworthy. Shackleton was distraught. Captain Sharpus, who worked hard on the ship in Rio, and observed Shackleton's moods, wrote:

> Poor Sir Ernest. He had a hell of a time here. Everything in the engine-room of that ship seemed to be wrong. He

> smiled to everyone, and to a certain extent made light of what was the matter. But I knew he was worried almost to death . . . I could see by his face it hurt him so. All his hopes and ambitions seemed centred in that little ship, and she seemed to let him down so much that I got to hate the damned boat. I was always glad when he came away from it. Often, when we left, if we were alone, he'd sit in the launch gazing at nothing, and hardly say a word on the half-hour run from the ship to the city.

Shackleton had hoped that a trip south with the old gang would see him rediscover his mojo, but so far it had failed to work its magic. 'There is something a little different in him this trip,' Macklin wrote, 'which I do not understand.' Unhappy at home and at sea, Shackleton cut a frustrated and downbeat figure, far removed from the forceful, gregarious personality of old. 'This is a lonely life after all,' he wrote to Emily, perhaps realizing that true happiness was destined always to evade him.

On 17 December he suddenly became short of breath. Macklin rushed to his side, fearing he had suffered a heart attack. Yet Shackleton brushed off all concern and refused to be examined, writing to Emily to say, 'Darling I am a little tired but all right. You are rather wonderful.'

When the *Quest* left Rio the next day it was obvious that Shackleton was not himself. Usually found prowling the decks, he was now confined to his cabin, finding any physical exertion beyond him. Macklin also noticed that he seemed to drink champagne in the morning, 'in the hope of staving off anginal attacks'. But those who had travelled with Shackleton before were well aware of his miraculous abilities to bounce back from illness and so did not as yet feel unduly concerned.

Due to all the delays, there was now no time to call at Cape Town to collect all the stores and gear. Shackleton instead hoped to get sufficient sledge equipment, dogs and clothing by returning to the scene of his greatest triumph: South Georgia. The stopover would also allow him to reminisce about past glories with the Norwegian whalers, many of whom remembered him fondly.

As ever, the trip to South Georgia was tough. The beleaguered *Quest* was lashed by ferocious storms for over a week, with Shackleton telling the men they were the worst he had ever experienced, which was certainly saying something. When it finally died down, the sea-sick men found that the main drinking water tanks had leaked and were virtually empty. A rigid water economy was enforced, and worse news followed. The ship's furnace had developed a leak, which reduced her speed. 'If this crack in the furnace proves serious I may have to abandon the expedition,' Shackleton confessed to his diary, then adding, 'I grow old and tired.'

One night, as Macklin was on watch with Shackleton, he became concerned at his behaviour. 'The Boss says now quite frankly that he does not know what he will do after South Georgia,' he wrote. 'I do not quite understand his enigmatical attitude – I wonder what we really shall do. Being in charge of equipment and stores, this uncertainty is very worrying.' Such indecision and pessimism were very unlike the bold and daring Shackleton of old.

Naisbitt, the acting steward, was also growing concerned, 'When the Boss is annoyed about something it seems to upset him altogether – nothing is right,' he wrote. 'At meal times there is something to complain about – the plates have not been warmed for the hot dishes – his macaroni cheese is not sufficiently crisp – or something else is wrong.' On previous expeditions, Shackleton had been happy to muck in and endure any hardship while putting a smile on everyone's face. Now his short temper betrayed the whirlwind of doubt and worry that was plaguing him inside.

The sight of the ice did at least briefly energize him. 'The old familiar sight aroused in me memories that the strenuous years had deadened,' he wrote. 'Ah me! the years that have gone since, in the pride of young manhood, I first went forth to the fight.' The return to South Georgia, on 4 January, was also a welcome tonic. Back on shore, he met up with some of his old friends and shared stories once more of his miraculous escape from *Endurance* six years before.

Reinvigorated, and with the dawn of a new expedition, an effervescent Shackleton boarded the *Quest* and wrote in his diary, 'A wonderful evening. In the darkening twilight I saw a lone star hover gem-like above the bay.'

That night of the 4–5 January 1922, he went to sleep happy and peaceful, finally ready for the task ahead and remembering why he so loved this way of life. However, at 2 a.m. he awoke with back pains. Upon being inspected by Macklin, he was given three aspirin and told to take it easy. He had clearly been overdoing things and had also been drinking far too much. Shackleton agreed and promised to rest, wanting to be in full health for the expedition ahead. Yet when Macklin returned an hour later, he found that no matter how hard he tried, he could not wake the man known to all as 'Boss'.

39

When dawn broke the next day, Wild gathered the crew around him. 'Boys, I've got some sad news for you,' he solemnly announced. 'The Boss died suddenly at three o'clock this morning. The expedition will carry on.' While there was shock and sadness, the crew recognized that it was exactly how Shackleton would have wanted to go. He had passed at the scene of his greatest triumph, and in the thrust of adventure, rather than fading away in an armchair dreaming of better days. As an admirer in the *Daily Mail* recalled him saying, 'He did not mean to die in Europe. He wanted some day to die away on one of his expeditions. "And," he was quoted as once saying, "I shall go on going, old man, till one day I shall not come back."'

It was no surprise to anyone that when Macklin performed a post-mortem and finally examined Shackleton's heart he found 'fatty extensive atheroma of the coronary arteries.' In short, Shackleton had died of heart failure at the age of forty-seven.

When Emily was told that her husband had died and arrangements were already well underway to return his body to London, she put her private wishes aside and said that his heart was always in the south, no more so than in South Georgia. She asked that he be buried there, where she knew he would be happiest and known to all as a hero. To the bitter end, Emily knew Shackleton better than anyone.

On 5 March 1922, Shackleton was duly buried at Grytviken's small wooden church in the shadows of the snow-capped mountains from which he had once climbed to salvation. Sadly, with the Shackleton–Rowett Expedition well underway, only Hussey could attend, with six ex-servicemen from the Shetland Islands carrying the coffin through the procession of Norwegian whalers, who all lowered their caps as a sign of respect.

A month later, with the *Quest* having completed various planned research tasks, she returned to Grytviken, where Wild and the others could at last see off the Boss. After kneeling at his old friend's grave, an emotional Wild wrote, 'I have not the least doubt that had Sir Ernest been able to decide upon his last resting place, it is just here that he would have chosen to lie.' Indeed, Wild would eventually join him. After dying in South Africa in 1939, his remains lay undiscovered until 2011, when he was then laid to rest next to Shackleton in Grytviken, a loyal companion until the very end.

In contrast to the simple and moving ceremony in Grytviken, Shackleton was given a truly worthy send-off back home. While he might have felt unappreciated during much of his life, the subsequent memorial service at St Paul's Cathedral was fitting for a hero in death. Emily and the children sat in the front row, with old schoolfriends, comrades from his various expeditions, representatives of royalty, the Admiralty, and even the Royal Geographical Society, all paying tribute to a man who had lived a remarkable life.

However, while the ceremony would have meant a great deal to Shackleton, he was a man who strived to be a legend. So, a century since his death, how has Shackleton been remembered?

At the time of his death, and for many decades thereafter, Shackleton's exploits were diminished in relation to those of his one time-rival Captain Scott, whose death so haunted the nation. While the Educational Council decreed that Scott's story was to be taught in schools, to some 750,000 schoolchildren, it was perhaps Fleet Street which was responsible for the disparity between the memory of the two rivals. The press chose to frame Scott not as a polar amateur beaten by a professional rival, nor as a great leader and explorer, but above all as a man who could inspire his countrymen to triumph over adversity, dying in order to reach his goal.

Shackleton certainly had plenty of failures under his belt, not once having reached his intended main goal, but crucially, he had not died a heroic and romantic death in the course of an expedition. Indeed, his greatest achievement, saving the men of the *Endurance*, was largely overshadowed by the war. At the time, the public were less interested in men who survived self-inflicted pain, striving for personal glory, whilst millions were dying fighting for their freedom on the front lines of Europe's battlefields. In the affections of the public, the war hero had now replaced the explorer hero. Decades of relative indifference to Shackleton followed, and he was almost consigned to long-forgotten history.

One story from the fifties, told by Shackleton's daughter Cecily, emphasizes just how far his star celebrity status had fallen. 'A short time ago the postman came here and said, "I see the name Shackleton. Are you by any chance related to the cricketer?" I said, "No, I'm afraid not." He said, "Oh, bad luck, I thought there was somebody connected with somebody interesting living here.'"

There would, however, be a remarkable reversal of fortunes for Shackleton's memory in the seventies, when South African journalist Roland Huntford ruthlessly attacked Captain Scott in his book *The Last Place on Earth*. I have gone on record to state that I found this book very inaccurate, not to mention, in my opinion, wrong. To my mind, Huntford engaged in a blatant character assassination, which was designed for nothing more than to sensationalize and sell books. The Scott that was portrayed in those pages was nothing like the one I have come to know through decades of research, as well as through walking in his footsteps. Sadly, this Huntford book, a clever mix of misinformation and disinformation, was widely acclaimed and greatly diminished the legend of Scott.

Huntford was far kinder to Shackleton in his follow-up biography, calling him the 'pre-eminent exponent of the so-called Heroic Age of Antarctic exploration'. At this, Shackleton's star rose once again, with an explosion of interest in his story, particularly that of the *Endurance* epic.

With his post-Huntford fame finally spreading beyond the relatively small community of polar scholars and enthusiasts, Shackleton's exploits captivated the imagination of a worldwide audience. Books, television documentaries and cinematic dramas told the story of his polar adventures and he especially became known for leadership under adversity.

His ability to motivate his men and put the interests of the group above all else has since been studied far and wide. Shackleton modules are taught at Harvard Business School, and the introduction of *Shackleton's Way: Leadership Lessons from the Great Antarctic Explorer* by Margot Morrell and Stephanie Capparell states that Shackleton 'has been called the greatest leader that ever came on God's earth, bar none . . .'

Such has been the rejuvenation in how Shackleton is remembered that in 2019 BBC viewers voted him as the nation's favourite polar icon, despite the fact that he never reached the pole. Far more than any of his actual achievements, it is the strength of his character that continues to reverberate through the ages.

So, what would Shackleton himself have made of all of this? I think it's fair to say he'd be pretty pleased. There is no doubt that it would be easy to pick apart aspects of his character and deem him a failure. To many, he failed in reaching his expedition goals, handling money and being an attentive husband and father. And yet, at least when it comes to each of his expeditions, you could also choose to rate them as a great success. He certainly helped pave the way for Amundsen and Scott to reach the South Pole. Above all, on the *Endurance*, he somehow kept all his men alive, despite a succession of the most perilous situations imaginable.

Some have suggested that Shackleton's plan to cross the Antarctic was doomed to failure and that he unnecessarily put his men's lives at risk. I recall discussing this subject with fellow veteran Sir Vivian Fuchs in the seventies. He was convinced that had Shackleton reached Vahsel Bay, he would still have failed, and almost certainly died in his attempt. In his opinion, Antarctica's summer season was simply too short to allow any human crossing of the mostly unexplored plateau between Vahsel Bay and Mount Hope. Shackleton would probably have faced a largely winter journey, through unknown territory with over 1,000 miles to travel to the first depot at Mount Hope, which the Ross Sea party of *Aurora* managed to lay. While Sir Vivian believed they would have certainly perished, I humbly disagree.

In 1993, when Mike Stroud and I man-hauled, unassisted, across the Antarctic continent, we took the same route as Shackleton intended to use. Before we set off, everyone told us it was impossible and that we were mad to even try, yet we achieved this feat, hauling a 485-lb start-load across 1,350 miles. Who knows what conditions Shackleton might have faced on his journey, but Shackleton and his team were undoubtedly younger and stronger than Mike and I were at the time, while the calorie usage and climatic conditions would have been similar. I hope our success proved that, despite what the critics might say, Shackleton's grand crossing plan was indeed possible, even if his preparations left a lot to be desired.

However, as we have seen throughout this book, merely conquering the south was in itself never truly Shackleton's overall dream. From a young age, he had regaled his sisters with tales of his courageous deeds, saving London from an inferno and being heralded as a hero. In life, he lived out this dream, heroically saving his men from the ice. When most others would have wilted under the pressure, he showed strength of leadership and astute decision-making which over a century on continue to impress. Perhaps Shackleton himself never truly realized it, but through the failure of *Endurance* he in fact achieved his ambitions to be a hero, with or without reaching the pole. His *Endurance/James Caird* voyage was undoubtedly the feat for which he will always be renowned, an example of extraordinary leadership under the most intense pressure, with lessons we can all learn from.

When it comes to summing up Shackleton, his polar contemporary Sir Raymond Priestley commented, 'For scientific leadership, give me Scott. For swift and efficient travel, Amundsen. But when you are in a hopeless situation, when there seems to be no way out, get on your knees and pray for Shackleton.' I could not have put it better myself.

Acknowledgements

My sincere thanks to James Leighton and to all at Penguin involved with producing this book, in particular Rowland White and Ariel Pakier, as well as my agent, Catherine Summerhayes. My thanks also go to the Royal Geographical Society and the Scott Polar Research Institute Archives and my many friends there; to my family for their patience at home; and to the Greenlands team, especially Jill Firman, for her tenacity and ability to translate the near-indecipherable. And to all those listed below, plus those I have stupidly omitted:

To Gill Poulter of the Dundee Heritage Trust, and all who have kindly helped with information at BAS, SPRI, and the RGS: Dr Julian Dowdeswell (former Director of SPRI), Professor Joe Smith (Director of the RGS), Ann Savours Shirley, Robert Headland (Institute Associate of SPRI), Peter Wadhams and Bob Burton (former Director of the South Georgia Museum) of BAS, and to Dr Peter Clarkson of SPRI.

Also to Anton Browning, the late Colonel Andrew Croft, the late Sir Vivian Fuchs, the late Dr John Heap, the late Sir Wally Herbert, Laurence Howell, Anton Bowring, Jill Bowring, Andrew Riley (Senior Archivist of the Churchill Archives Centre), Alastair and Eugene at the RGS Archives, Nigel Winser, Professor Mike Stroud, Oliver Shepard, the late Charles Swithinbank, Brigadier Mike Wingate Gray and all at SAS Group HQ; Jonathan Shackleton, Rear Admiral Jeremy Larken (MD of OCTO); and to Brad Borkan.

Bibliography

Alexander, Caroline, *The Endurance*, Bloomsbury, 1999
—, *Mrs Chippy's Last Expedition*, Bloomsbury, 1997
Amundsen, Roald, *The South Pole*, John Murray, 1912
Armitage, Albert B., *Two Years in the Antarctic*, Bluntisham Books, 1984
Bernacchi, L.C., *Saga of the 'Discovery'*, Blackie & Son, 1928
Bickel, Lennard, *Mawson's Will*, Stein & Day, 1977
—, *Shackleton's Forgotten Men*, Adrenaline Classics, 2000
Borkan, Brad and Hirzel, David, *When Your Life Depends on It*, Terra Nova Press, 2017
Brown, Paul, *The Last Wilderness*, Hutchinson, 1991
Browning, Robert, *Poetical Works*, Oxford University Press, 1970
Cameron, Ian, *Antarctica: The Last Continent*, Cassell, 1974
Cherry-Garrard, Apsley, *The Worst Journey in the World* (first pub. 1922), Chatto & Windus, 1951; Picador, 1994
Churchill, Winston, *The Great War*, George Newnes Ltd, 1933
Davis, J. K., *His Antarctic Journals*, Melbourne, 1962
Doorly, Gerald, *The Voyages of the 'Morning'*, Smith Elder, 1916
Dowdeswell, Julian and Hambrey, Michael, *The Continent of Antarctica*, Papadakis, 2018
Dunnett, H. M., *Shackleton's Boat*, Neville & Harding, 1996
Evans, Edward (Mountevans), *The Antarctic Challenged*, Staples Press, 1955
Fiennes, Ranulph, *Beyond the Limits*, Little, Brown, 2000
—, *Captain Scott*, Hodder & Stoughton, 1993
—, *Colder*, Simon & Schuster, 2016
—, *Hell on Ice*, Hodder & Stoughton, 1979
—, *Icefall in Norway*, Hodder & Stoughton, 1972

—, *Mind over Matter*, Sinclair-Stevenson, 1993
—, *To the Ends of the Earth*, Hodder & Stoughton, 1983
Fisher, M. and J., *Shackleton*, Barrie Books, 1957
Fuchs, Sir Vivian, *Of Ice and Men*, Anthony Nelson, 1982
—, *A Time to Speak*, Anthony Nelson, 1990
Fuchs, Sir Vivian and Hillary, Sir Edmund, *The Crossing of Antarctica*, Cassell, 1958
Giaever, John, *The White Desert*, Chatto & Windus, 1954
Gran, Tryggve, *The Norwegian with Scott* (first pub. 1915 in Norwegian), HMSO, 1984
Headland, Robert, *The Island of South Georgia*, Cambridge University Press, 1984
Herbert, Wally, *A World of Men*, Eyre & Spottiswoode, 1968
Huntford, Roland, *Scott and Amundsen*, Hodder & Stoughton, 1979
—, *Shackleton*, Atheneum, 1986
Hurley, F., *Argonauts of the South*, Putnam, 1925
Hussey, L. D. A., *South with Shackleton*, Sampson Low, 1949
Huxley, Elspeth, *Scott of the Antarctic*, Weidenfeld & Nicolson, 1977
John, Brian S., *The Ice Age*, Collins, 1977
Joyce, E. M., *Correspondence*, SPRI, n.d.
Keir, David, *The Bowring Story*, Bodley Head, 1962
Lansing, Alfred, *Endurance*, Hodder & Stoughton, 1959
Law, Phillip, *Antarctic Odyssey*, Heinemann, 1983
Lees, T. O. H., *Antarctic Journal*, Turnbull Library, n.d.
Limb, Sue and Cordingley, Patrick, *Captain Oates: Soldier and Explorer*, Leo Cooper, 1982
Mackay, A. Forbes, *Nimrod Diary*, SPRI, 1909
Markham, Sir Clements, 'The Antarctic Expeditions', *Geographical Journal* vol. XIV, no. 5 (1899)
Marshall, E. S., *Nimrod Diaries 1907–1908*, RGS, n.d.
Mason, Theodore K., *On the Ice in Antarctica*, Dodd, Mead, 1978
Mawson, Sir Douglas, *The Home of the Blizzard*, Heinemann, 1915

McKenna, John and Shackleton, Jonathan, *An Irishman in Antarctica*, Lilliput press, 2003
Mear, Roger and Swan, Robert, *In the Footsteps of Scott*, Jonathan Cape, 1987
Messner, Reinhold, *Antarctica: Both Heaven and Hell*, Crowood Press, 1991
Mill, H. R., *The Life of Sir Ernest Shackleton*, William Heinemann, 1923
Mills, Leif, *Frank Wild*, Caedmon of Whitby, 1999
—, *Men of Ice*, Caedmon of Whitby, 2008
Mitchener, E. A. (Ted), *Ice in the Rigging*, Southerly Press, 2008
Morrell, M. and Capparell, S., *Shackleton's Way*, Nicholas Brealey Publishing, 2003
National Maritime Museum, *South: The Race to the Pole*, National Maritime Museum, 2000
Partridge, Bellamy, *Amundsen*, Robert Hale, 1953
Peat, Neville, *Shackleton's Whisky*, Preface Publishing, 2013
Richards, R. W., *The Ross Sea Shore Party, 1914–17*, SPRI, 1962
Riffenburgh, Beau, *The Myth of the Explorer*, Belhaven Press, 1993
—, *Nimrod*, Bloomsbury Publishing, 2005
Royds, Charles, *Diary*, T. Roger Royds, 2001
Sarolea, C., 'Sir Ernest Shackleton: A Study in Personality', *Contemporary Review*, vol. 121 (1922)
Savours, Ann, *The Voyages of the Discovery*, Virgin Books, 1992
Scott, Robert Falcon, *The Diaries of Captain Robert Scott*, University Microfilms, 1968
—, *Scott's Last Expedition*, Smith, Elder, 1913
—, *Scott's Last Expedition: The Personal Journals of Capt. R. F. Scott CVO RN*, John Murray, 1973
—, *The Voyage of the 'Discovery'*, 2 vols, Smith, Elder, 1905
—, *The Voyage of the 'Discovery'*, 2 vols, Macmillan, 1905
Seaver, George, *Edward Wilson of the Antarctic*, John Murray, 1933
—, *The Faith of Edward Wilson*, John Murray, 1948

Shackleton, Ernest, *The Heart of the Antarctic*, Heinemann, 1932
—, *Letters to Emily Shackleton*, SPRI, n.d.
—, *South*, Heinemann, 1919
Smith, Michael, *Shackleton: By Endurance We Conquer*, Oneworld Publications, 2014
—, *An Unsung Hero: Tom Crean – Antarctic Survivor*, Collins Press, 2000
Solomon, Susan, *The Coldest March*, Yale University Press, 2001
Steger, Will and Bowermaster, Jon, *Crossing Antarctica*, Bantam Press, 1991
Strachey, Lytton, *Eminent Victorians*, Penguin, 1986
Stroud, Mike, *Shadows on the Wasteland*, Penguin, 1993
—, *Survival of the Fittest*, Jonathan Cape, 1998
Sugden, David, *Arctic and Antarctic*, Blackwell, 1982
Swithinbank, Charles, *An Alien in Antarctica*, McDonald and Woodward Publishing Company, 1997
Tyler-Lewis, Kelly, *The Lost Men*, Bloomsbury, 2006
Wheeler, Sara, *Terra Incognita*, Jonathan Cape, 1996
Wild, Frank, *Shackleton's Last Voyage*, Cassell, 1923
Wilson, D. M. and Elder, D. B., *Cheltenham in Antarctica*, Reardon Publishing, 2000
Wilson, Edward, *Diary of the Discovery Expedition to the Antarctic Regions 1901–1904*, Humanities Press, 1967
—, *Diary of the Discovery Expedition to the Antarctic Regions 1901–1904*, ed. Ann Savours, Blandford Press, 1966
Worsley, Frank A., *Endurance*, W. W. Norton, 1931
—, *Shackleton's Boat Journey*, Folio Society, 1974

Picture Credits

Courtesy of Ran Fiennes, p.4: 12b; p.7: 21b; p.8: 23b; p.10: 25b, 26; p.11: 27; p.13: 33b; p.15: 39
© C. R. Ford/Royal Geographical Society via Getty Images, p.2: 5
© Christies/Bridgeman Images, p.2: 7
© Ernest Shackleton/Royal Geographical Society via Getty Images, p.4: 12c; p.5: 13; p.6: 16
© Frank Hurley/Royal Geographical Society via Getty Images, p.4: 12a; p.7: 20, 21a; p.8: 22, 23a; p.9: 24a, b, d; p.10: 25a; p.11: 28, 29; p.12: 30, 31a & b; p.14: 35
© Illustrated London News, p.5: 15
© John Thomson/Royal Geographical Society via Getty Images, p.3: 8
© Mrs W. Parke/Royal Geographical Society via Getty Images, p.14: 34
© Royal Geographical Society via Getty Images, p.3: 11; p.4: 12d; p.5: 14; p.9: 24c; p.13: 32; p.16: 41
© Scott Polar Research Institute, University of Cambridge, p.6: 18
Unknown, p.1: 1–4; p.2: 6; p.3: 9, 10; p.6: 17–19; p.13: 33a, p.14: 36, 37; p.15: 38, 40

Every effort has been made to trace copyright holders and to obtain their permission for the use of copyright material. The publisher apologizes for any errors or omissions and would be grateful to be notified of any corrections that should be incorporated in future editions of this book.

Index

POLAR SHIP "ENDURANCE."

Built by the Framnæs Works, Sandefjord, Norway, and Purchased by the Imperial Transantarctic Expedition. (*See page* 126.)

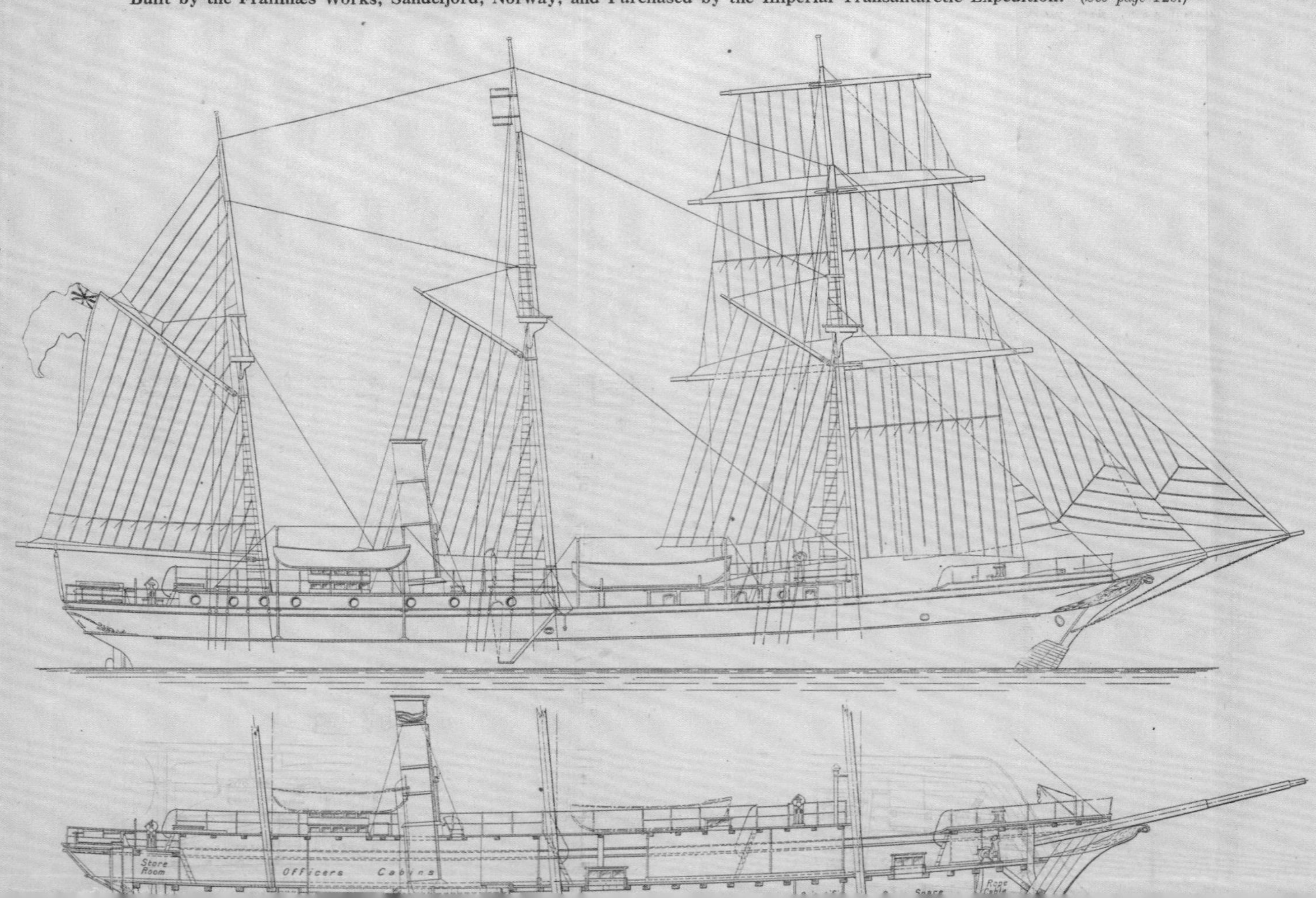